AF618919

Nico Bauer

What are the potential causes of incorrect predictions of tropical cyclone intensification in medium-range ensemble forecasts?

Bauer, Nico: What are the potential causes of incorrect predictions of tropical cyclone intensification in medium-range ensemble forecasts?, Hamburg, Bachelor + Master Publishing 2022
Originaltitel der Abschlussarbeit: What are the potential causes of incorrect predictions of tropical cyclone intensification in medium-range ensemble forecasts?

Buch-ISBN: 978-3-95993-113-7
PDF-eBook-ISBN: 978-3-95993-613-2
Druck/Herstellung: Bachelor + Master Publishing, Hamburg, 2022
Zugl. Karlsruhe Institut für Technologie, Deutschland, Masterarbeit, 2021

Bibliografische Information der Deutschen Nationalbibliothek:
Die Deutsche Nationalbibliothek verzeichnet diese Publikation in der Deutschen Nationalbibliografie; detaillierte bibliografische Daten sind im Internet über http://dnb.d-nb.de abrufbar.

Hermannstal 119k, 22119 Hamburg
http://www.bachelor-master-publishing.de, Hamburg 2022
Printed in Germany

Contents

1 Introduction

Tropical cyclones (TCs) are the most destructive weather phenomena in the world (Longshore, 2008). These storm systems are common in large parts of highly populated tropics and subtropics with favorable atmospheric conditions. Society's vulnerability to them and the associated annual economic costs have risen steadily Bevere et al. (2020): mean worldwide insured losses averaged 75 billion USD per year in the 10 years between 2009 and 2019 (Bevere et al., 2020). Cinco et al. (2016) analyzed TC data and observed that in the period from 1951 to 2013, an average of 19.4 TCs entered the Philippine Area of Responsibility in the Western North Pacific (WNP), and nine TCs moved over the islands. Consequently, the Philippines have the highest number of landfalling storms and the highest rate of severe TC worldwide. In turn, the TCs that move over the islands into the South China Sea (SCS) frequently affect the coast of Vietnam. Through a spatial assessment of TC vulnerability, Nguyen et al. (2019) have demonstrated a high or very high susceptibility in most parts of coastal Vietnam. The most extreme event over the WNP in the last century was Typhoon Haiyan, which caused 6,300 deaths and widespread economic and socioeconomic damage (Lagmay et al., 2015). This significant susceptibility implies the high importance of improving weather forecast models for greater predictive capability. In recent decades, the quality of forecasting tropical cyclone tracks has increased steadily at the European Center for Medium-Range Weather Forecast (ECMWF) and other numerical weather prediction centers (Yamaguchi et al., 2017). Nevertheless, intensity predictions still present more significant challenges (Rappaport et al., 2009; Cangialosi and Franklin, 2012; Bonavita et al., 2017).

Cyclones that affect the coasts of Southeast Asia are additionally influenced in their intensity by long-lasting cold surges from the Asian Continent (Takahashi et al., 2011). According to Chang et al. (2003), tropical cyclones that frequently develop in October or November over the WNP are strongly affected by the northerly wind surges over the SCS. In autumn, at the time of cold surge occurrence over the SCS, TCs formed over the WNP indicate a higher probability of straight westward movement over the Philippines than in other seasons (McBride, 1995). In summer, TCs more frequently demonstrate a recurvature to the north close to the Philippines. Therefore, the frequent cold surge occurrence period exhibits the highest TC impact for the Philippines and Vietnam. The induced pressure difference between the wind surge area outside the cyclone and the center leads to externally forced convergence in association with enhanced water vapor fluxes towards the storm center (Gray, 1998). Therefore, sub-monthly intense cold surges over the SCS enhance convective activity there (Compo et al., 1999). This convective activity is due to the higher amount of latent heat release that is an essential contributor to tropical cyclone intensification (TCI) (McBride, 1995).

The lower quality of intensity forecasts are mainly due to small-scale processes like random, chaotic moist convection at the beginning of a rapid intensification stage Tao and Zhang (2014) and differences in dynamical parameters, such as vertical wind shear in regimes with a strongly vertical and horizontal fluctuating wind field (Komaromi and Majumdar, 2014). These limitations are enhanced under increasing environmental vertical wind shear conditions, especially for onset time of rapid intensification (Zhang and Tao, 2013). Komaromi and Majumdar (2014) used several metrics to investigate quantification of predictability using the Ensemble Prediction System (EPS) at ECMWF. The study shows that 850–200 hPa bulk environmental vertical wind shear is predictable one week prior, while 200 hPa divergence and 850 hPa vorticity are only predictable 24 hours prior. Komaromi and Majumdar (2015) also indicate a strong sensitivity of relative humidity at 700 hPa at the beginning and the end of the cyclogenesis due to the timing and location of dry air in the middle troposphere. Values under 60 % are within a radius of 300 km from the circulation center, a limiting factor for genesis. According to Emanuel (1986), deep column moisture and convective available potential energy cause much of the ensemble spread. These uncertainties are worsened by differences in convection-related latent heat fluxes and wind-induced surface heat exchange (WISHE) processes. These discrepancies cause minor moisture discrepancies in the boundary layer that are highly influential on moist convection (Van Sang et al., 2008). As a result, intensity forecasts are intrinsically limited.

In this thesis, TC track data from the The Observing system Research and Predictability Experiment (THORPEX) THORPEX Interactive Grand Global Ensemble (TIGGE) database from 2011 to 2020 is used to identify three case studies with the most substantial spread in predicting TC intensification. Vitart et al. (2012) used the mean absolute intensity and position error to compare the skill of intensity forecasts for two different cyclone tracking systems. A mean intensity error is computed in this study to determine whether a TC forecast is defined as an over-forecasting (OFC) event with an enormously increasing negative value or an under-forecasting (UFC) with a significantly growing positive value. Therefore, an OFC is a tropical cyclone event with a rapidly rising negative mean error in a forecast period up to 120 h. Additionally, the spread in the ensemble is considered to hold a clear separation in the intensity of the different clusters. Hamill et al. (2011) computed the mean spread for Hurricane Sandy in the global forecasting system ensemble forecast to analyze their track evolution. In this work, the formula is applied as the mean spread of intensity.

Gaining a better understanding of the processes behind this prediction error is the main target of this study. For this purpose, three events with the strongest OFC and UFC and the highest divergence in the ensemble after 120 h are selected. A k-mean clustering according to minimum pressure and location after lead time is used to organize the members and make them comparable (Hartigan and Wong, 1979). The calculation of normalized differences provides a better comparison between the different determined clusters of the atmospheric state variables at various vertical levels, fields, and times (Torn et al., 2015). The ventilation index Tang and Emanuel (2012) is computed to describe the thermodynamic and dynamic processes that are subject to tremendous possible uncertainties. This includes the relationship between the bulk environmental vertical wind shear, the maximum potential intensity Bister and Emanuel (2002), and the entropy deficit (Tang and Emanuel, 2010). Additionally, the calculation of the modified cold surge index from Chang et al. (2005) expresses the sensitivity in the model regarding enhanced moisture fluxes and convective activity. Contributions to intensification by the corresponding latent heat release are evaluated by the (Fink et al., 2012, their pressure tendency equation (PTE)). This differential equation includes five different terms and relates their amounts to the pressure fall of an extratropical cyclone. In this study, the PTE is applied to storms with a warm core thermal structure for the first time.

The theoretical background of favorable atmospheric conditions, genesis, and indices used in Chapters 5 and 6 are explained in Chapter 2. Chapter 3 defines the research questions. The used data and applied methods are stated in Chapter 4. In Chapter 5, the case studies with over-forecasting are analyzed regarding their sensitive acting environmental conditions. The strongest under-forecasting event is studied in Chapter 6, followed by a discussion of the discrepancies between the different locations and case studies. Chapter 7 analyzes the findings from the PTE and their implications for predicting tropical cyclone intensity. In Chapter 8, the main points are summarized and discussed, recording results from previous studies, along with a brief outlook for future research.

2 State of Research

The first part of this chapter describes the general characteristics of TCs and their environmental and synoptical features that affect the development and intensification process. In the second part, their distinctions across different ocean basins are explained. Various uncertainty causes from ECMWF are characterized in (Sec.2.2), along with a comparison between the parts within the WNP since this is the central area of the case studies. This is followed by an introduction into forecast uncertainties in the EPS relating to TCI. Finally, indices for studies of primary error source-related variables are described based on their application reasons. Additionally, the important principle of convective heating in TCI is illustrated.

2.1 Tropical cyclones

TCs are defined as non-frontal synoptical low-pressure systems over warm water in the tropics or subtropics with a closed cyclonic surface wind circulation that includes organized deep convection (McBride, 1995). Maximum sustained wind speed and minimum central pressure are the main characteristics of their intensity. The maximum sustained wind speed over 1 or 10 min (dependent on their location) is commonly used to define different intensity stages (Longshore, 2008). The threshold value between a tropical depression and a tropical storm is $17\,\mathrm{ms}^{-1}$. If wind speeds are between 18 and $32\,\mathrm{m\,s}^{-1}$, they are classified as tropical storms. In contrast, storms with wind speeds above $32\,\mathrm{m\,s}^{-1}$ are called hurricanes in the Western North Atlantic or Eastern North Pacific, typhoons in the study area of the WNP, and severe tropical cyclones elsewhere (Emanuel, 2003).

The circulation of a mature TC with wind speed above $32\,\mathrm{m\,s}^{-1}$ is mainly characterized by its primary and secondary circulation. The primary circulation is equivalent to the tangential wind of a cyclone, which is given as a first approximation in gradient wind balance above the frictional forced boundary layer McBride (1995), defined as:

$$v_{\mathrm{G}} = \frac{v_{\mathrm{c}}}{\frac{1}{2R_{\mathrm{c}}} + \sqrt{1 + \frac{1}{4R_{\mathrm{c}}^{2}}}}, \tag{2.1}$$

where v_{c} is the cyclostrophic wind, which is approximately the wind in the inner core of the cyclone and

$$R_{\mathrm{c}} = \frac{v_{\mathrm{c}}}{f_{\mathrm{r}}} \tag{2.2}$$

is the Rossby number with the Coriolis parameter f_{r}. R_{c} is large near the cyclone center and small in the outer areas of the TC, where the Coriolis force dominates and the winds are roughly in geostrophic balance. Several investigations have shown that the strength of the primary circulation of a cyclone is proportional to the intensity of its vortex (Gray, 1998; McBride, 1995; Emanuel, 2003). Therefore, it is of particular importance to consider the physical principle of the conservation of angular momentum. The angular momentum equation

$$M = rv + \frac{f}{r^2} \tag{2.3}$$

is considered to give a deeper understanding of the spin-up mechanisms of the vortex. In (Eq.2.3), is r the radius and v the wind speed. If an air parcel moves to a smaller radius while conserving M, the speed v increases according to the conservation of angular momentum. As a result, the highest wind speeds in TCs occur in the eyewall close to the center.

The secondary circulation comprises the radial and vertical components of the flow.

$$\frac{\partial p}{\partial r} = \rho(\frac{v^2}{r} + fv) \tag{2.4}$$

In the vertical the circulation is forced by latent heat release in deep convective clouds and friction. Scale analysis shows that the gradient wind balance is not valid in the boundary layer (Smith and Montgomery, 2016b). In the lowest layer, the friction-induced and intense inward-directed flux of water vapor converges. Consequently, the air rises rapidly. The release of latent heat overcompensates the downward motions of air parcels by evaporation and fosters the secondary circulation (Longshore, 2008, Fig.2.1).

Before a TC shows a closed circulation system, a cluster of convective clouds transforms into a tropical storm. This process is called tropical cyclone genesis (TCG). According to Gray (1968), TCG represents the organization of random and chaotic convective cells into a mesoscale vortex under favorable environmental conditions. In contrast, Nolan et al. (2007) used a wind speed threshold of $20\,\mathrm{m\,s^{-1}}$ to determine the point of genesis. Earlier studies (Zehr (1992); Emanuel (1993)) divide the formation process into two or more stages with a transition point at the formation of a mesoscale vortex. A mesoscale vortex emerges from a long-lived convective weather system called a mesoscale convective system (Houze et al., 1981). This convective system is identified as a large area with cloud top temperatures lower than $-70°$C, which signifies the presence of a long-lived altostratus deck. Initial cloud clusters may contain one or more mesoscale convective systems. During stage 1 of the genesis process, a mesoscale vortex is embedded within an existing cloud cluster; in stage 2, the central pressure decreases and the tangential wind strengthens, resulting in a TC (Fig.2.2).

Previous studies have underlined the importance of the marsupial paradigm to the formation process of TC (Dunkerton et al., 2009; Lussier III et al., 2014). This theory demonstrates that the critical layer of a tropical easterly wave plays a crucial role in TC formation. A critical wave layer is developed from a wave breaking or roll-up of cyclonic vorticity in the lower atmosphere, providing a favorite area for vorticity seedlings. In the north Atlantic, the initial wave is primarily the African easterly wave (Russell et al., 2017). In contrast, in the WNP, this process often develops in a trough of the Pacific easterly wave (Lussier III et al., 2014). A closed circulation with rapid moistening by

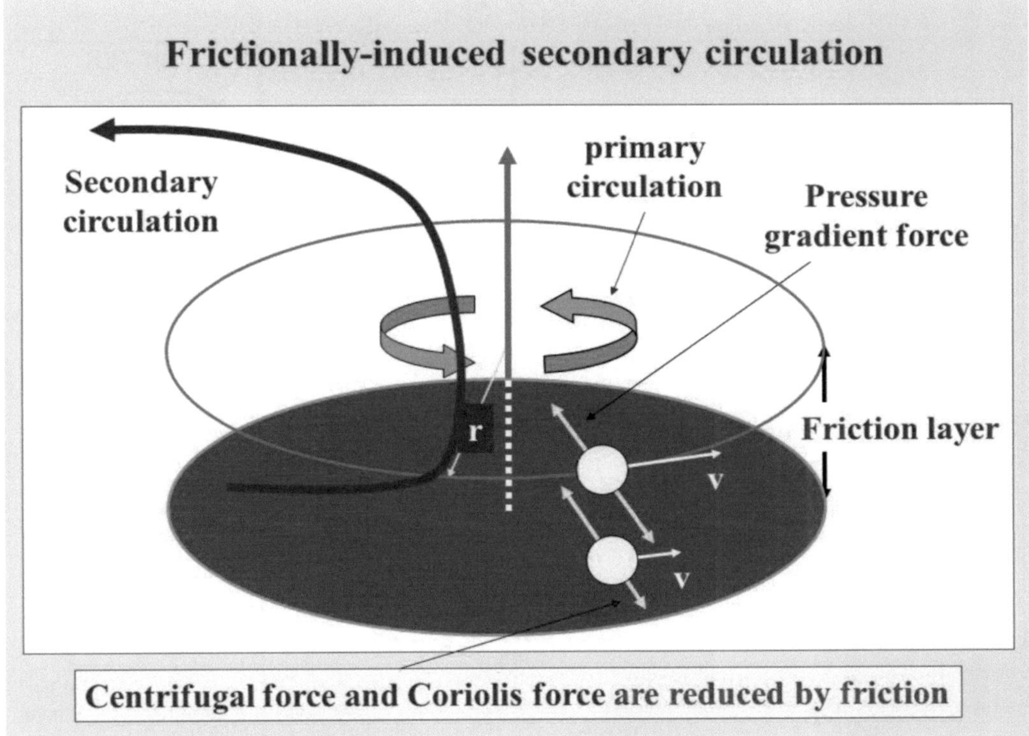

Figure 2.1: Schematic illustration of the circulation systems in an mature TC. The green circle around the vertical central axis describes the primary circulation. The blue arrow demonstrates the secondary circulation. The different forcing mechanisms of the two circulation systems are also demonstrated (Smith and Montgomery, 2016b).

convective activity and protection from dry air around is formed in this area (Fig.2.3). The parent wave pouch, which is more resistant against the surrounding dry air, is sustained and possibly enhanced by diabatically amplified vortices within the wave. As a result, the convective system develops a stable vortex and maintains a closed circulation.

2.1.1 Favorable environmental conditions

Several environmental conditions are necessary for the development of a TC from an initial convective cluster. According to Gray (1975, 1977) , there are six different parameters for seasonal TC formation preferable to TCG:

1. Significant Coriolis parameter f
2. Strong relative vorticity ζ
3. Low deep-layer vertical wind shear
4. Sea surface temperatures above 26°C to a depth of 60 meters
5. Clear differences in equivalent potential temperature between surface and mid-troposphere (500 hPa) $\Delta\Theta_e$

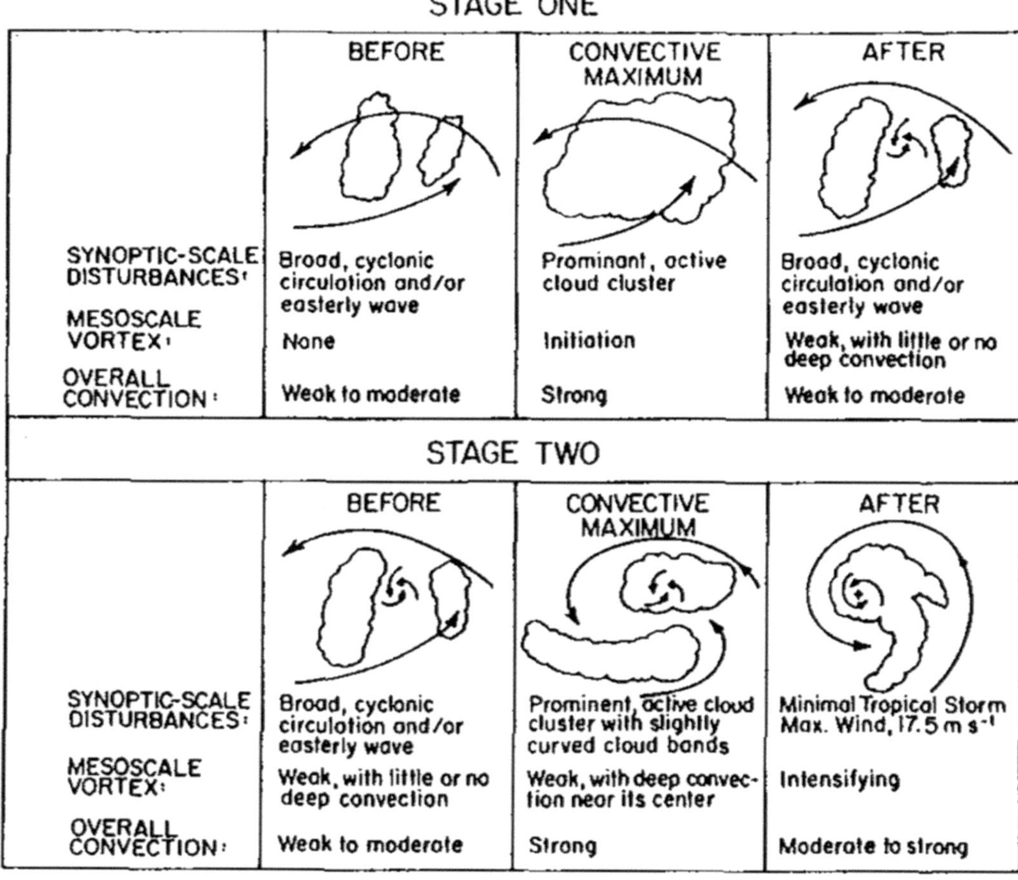

Figure 2.2: Schematic illustration of tropical cyclogenesis with using characteristics that are observable using via satellite imagery. The formation process is separated into two stages that describe the development from a convective system with no mesoscale vortex to a tropical cyclone with an intensifying mesoscale vortex and spiral convective cloud clusters (Gray, 1998).

6. High relative humidity in the midlevel.

A non-negligible Coriolis parameter favors cyclonical circulation. No tropical cyclone can exist directly at the equator (or close to it) due to the decrease of the Coriolis force to zero toward the equator. Therefore, storms only occur at latitudes above 5° to the north and south of the equator (McBride, 1995).

Large atmospheric values of low-level vorticity produce a more substantial frictionally forced convergence of mass and water vapor. This flux causes upward motion in the inner core of the storm and enhances the secondary circulation (Gray, 1968, Sec.2.1). Additionally, the strength of low-level vorticity corresponds to the strength of the primary circulation, which describes the intensity of the tangential winds (Gray, 1998). Therefore, higher values of initially relative vorticity from clusters of thunderstorms or mesoscale convective systems favor TCI.

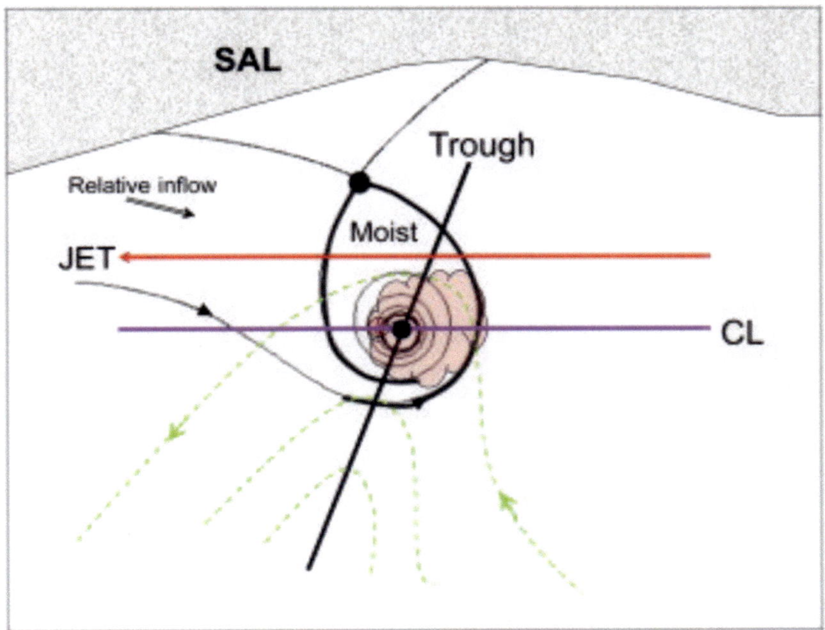

Figure 2.3: Formation of a TC in a wave pouch according to the marsupial paradigm. The solid line marks the wave pouch. The dashed contours represent the streamlines that are influenced by the wave structure. The pouch can protect the mesoscale vortices from the hostile environment, such as dry air associated with the Saharan air layer. Deep convection (gray shading) is sustained within the pouch. The easterly jet, the critical latitude, and the wave trough axis (Trough) are additionally indicated in the schematic. (Dunkerton et al., 2009).

Gray (1968) characterized the deep-layer vertical wind shear $|V_z|$ as the magnitude of the horizontal wind between the lower troposphere at 850 hPa and the upper troposphere at 200 hPa. Several numerical studies indicate its crucial role in regulating TCI (Goldenberg et al., 2001; Tang and Emanuel, 2010; Tao and Zhang, 2014). Tao and Zhang (2014) show the detrimental effect of strong vertical wind shear on convection and the influence on its spatial distribution. This process causes negative feedback between diabatic heating, the primary circulation, and the strength of the cyclone (Sec.2.3.4). Two factors drive this detrimental effect. First, a strong surrounding flow leads to a vertical slope alignment of the axes between the low- and mid-level vortex. As a consequence, the released heat is no longer vertically stacked, and intensity decreases (Fink and Speth, 1998). Second, the declined axis enhances the probability of intrusion of dry air masses from the midlevel into the low-level central area of the TC (Tang and Emanuel, 2010). The ventilation index, according to Tang and Emanuel (2012), describes this process (Sec.2.3.1). Based on these processes, thresholds are derived to estimate the occurrence of TCG and TCI. According to Goldenberg et al. (2001), values above $8\,\mathrm{m\,s^{-1}}$ are unfavorable for TC development, and values over $15\,\mathrm{m\,s^{-1}}$ are highly unfavorable for TCI. In contrast, Shu et al. (2013) indicate a critical value for TC formation of $10\,\mathrm{m\,s^{-1}}$ over the WNP. Additionally, it is stated that for TCI, the low layer vertical wind shear between 850 hPa and 10 m is more important, and $1.5\,\mathrm{m\,s^{-1}}$ is identified as the critical value.

The decreasing effect on TC intensity of an enhanced vertical wind shear depends on the moisture content in the mid-troposphere. Frank and Ritchie (1999) used numerical simulations of tropical cyclone-like vortices to examine the impact of a dry and wet vortex on unidirectional wind shear. Their simulations illustrate the vortex's higher resistance and intensification rate in the moist simulation than the dry one, which fails to progress and weakens. According to Gray (1968) , a relative humidity under 40 % is an inhibiting factor of TC formation. Similarly, Ge et al. (2013) show that the relative location of air with a low humidity to the right side of the vertical wind shear direction is especially detrimental to a TC since the dry air is advected by the circulation to the down-shear side and taken into the updraft. High saturation fractions in the inner core are the initiators of sustained deep convection. Wang and Hankes (2016) identified a non-linear relationship between saturation fraction and precipitation rate in tropical disturbances. Numerical simulations show a strong moistening in the inner region of an initial cluster of thunderstorms two or three days before genesis, which leads to a substantial increase in precipitation and low-level vorticity (Wang, 2012; Helms and Hart, 2015; Wang and Hankes, 2016).

The amount of moisture in the lower troposphere is strongly dependent on the stored amount of heat in the ocean (McBride, 1995). Therefore, the sea surface temperature is the primary energy source in a TC, providing latent and sensible heat input. 26°C is the threshold for development and further intensification over vast parts of ocean basins (Longshore, 2008). Conversely, Wang et al. (2007) show that 27°C is a prerequisite for tropical cyclones over the SCS and 28°C for their rapid intensification. A rapid intensification stage is defined as a wind increase by $15{,}4\,\mathrm{m\,s^{-1}}$ or greater within a 24-hour period. Fudeyasu et al. (2018) analyzed TCs in the WNP and found that the most rapid intensification events occur for sea surface temperatures higher than 28,4°C. As a result, more heat over the SCS is essential to intensify a TC than over parts of the WNP.

The transformation of heat from the ocean into kinetic energy is controlled by the vertical mass flux from the ocean surface (Murthy and Boos, 2018). A large surface enthalpy flux near the circulation center of a TC results from two mechanisms. First, increased surface wind enhances surface enthalpy fluxes, which contribute to a negative radial gradient from the center to the inner core area. Second, a constant surface wind speed leads to air-sea enthalpy disequilibrium (Murthy and Boos, 2018). In turn, the increase in fluxes is locally produced (Gao et al., 2019). In contrast, Fritz and Wang (2013) identify as the primary contributor, the inward moisture flux from outer areas of the storm for the radial gradient of enthalpy and corresponding increased precipitation rates. Additionally, Murthy and Boos (2018) suggest that a stronger radial gradient in the surface enthalpy fluxes enhances convective instability near the storm center.

2.1.2 Genesis factors

Early investigations assumed that TCG arises from preexisting disturbances in the lower troposphere, such as fronts or the upper troposphere (Riehl, 1948). Since then, studies and observations have highlighted many triggers for TCG dependent on ocean basins and seasonality, including classical baroclinic development, interaction with easterly waves or other low-level disturbances with upper-tropospheric disturbances, and accumulation of wave energy in large-scale diffluent flow (Emanuel, 2003). This study focuses mainly on TC genesis processes in the WNP and the SCS.

According to Ritchie and Holland (1999), large-scale flow patterns over the WNP, which are relevant for TCG, can be classified based on five factors (*monsoon shear line, easterly wave, monsoon gyre, rossby energy dispersion, monsoon confluence zone*, Fig.2.4). The monsoon shear line acts as a precursor synoptic-scale disturbance annually. Its location in the WNP is more than two the west, close to the Philippines (Fudeyasu et al., 2018). This is then enhanced by horizontal cyclonic shear associated with the monsoon trough, which extends from the Asian continent over the Philippines into the WNP, accompanied by the southwesterly flow on its southern side and easterlies on its northern side (Gray, 1998). Due to the small landmasses involved, only slight seasonal variations in the location of the monsoon shear line are observed annually. Yoshida and Ishikawa (2013) used reanalysis data to identify flow patterns for the genesis of 908 TCs between 1979 and 2008: 48 % were monsoon shear line-related cyclones, the most common type in this geographical area.

In contrast to easterly wave-forming cyclones, monsoon shear line storms show higher values of low- and mid-tropospheric humidity (Fudeyasu et al., 2018). This difference is due to their more northward and eastward genesis locations. As a result, they develop over colder water with weaker heat fluxes. These storms are formed in a developing Pacific easterly wave in association with the synoptic-scale easterly trade wind system (Li and Hsu, 2018). The Pacific easterly waves obtain energy from the mid-latitude jet stream, which shows the critical connection to extratropical storm systems in the mid-latitudes. The troughs within the easterly wave provide a favorite environment for these cyclones. Easterly wave storms indicate a smaller average size and have a more westward track than other types (Fudeyasu et al., 2018). Consequently, they more frequently influence the Philippines, the SCS and Vietnam (Yoshida and Ishikawa, 2013).

Gyre-related cyclones and storms developing in a confluence zone are associated with monsoon disturbances. The former emerge directly from a synoptic-scale gyre embedded within a developing monsoon trough, while the latter form in the zonal confluence zone between easterly trade winds and westerly monsoon winds on the eastern side of the monsoon trough (Li and Hsu, 2018). Confluence area storms show the highest intensification rate of all five storm types, while gyre-related cyclones indicate the slowest rate (Fudeyasu et al., 2018). The main reason for the difference in intensification rates between the two types is their distinct size. While the storms in a synoptic-scale gyre are the largest off all five storm types, the cyclones with the highest intensification rates are much smaller. Carrasco et al. (2014) confirm this result by showing an inverse relationship between storm size and rapid intensification rate during TC development.

The formation of a cyclone behind an existing intense storm through energy dispersion is the last genesis type. In this case, a mature TC disperses its energy as a Rossby wave train to the southeast; in one trough, a low-pressure system forms, which sometimes develops into a TC. An investigation by Ge et al. (2008) into 3D Rossby wave energy dispersion shows that the Rossby wave train has a baroclinic structure whose upper-level part develops quickly, triggering downward energy propagation that fosters the low-level wave train. This kind of storm genesis is restricted to the warmest water areas with a very high tropical cyclone heat potential, including a deep 26° C isotherm and huge heat content in the 100 m deep surface ocean layer (Fudeyasu et al., 2018). According to Yoshida and Ishikawa (2013), TCs are often not only developed by one of these factors but through a combination of multiple flow patterns.

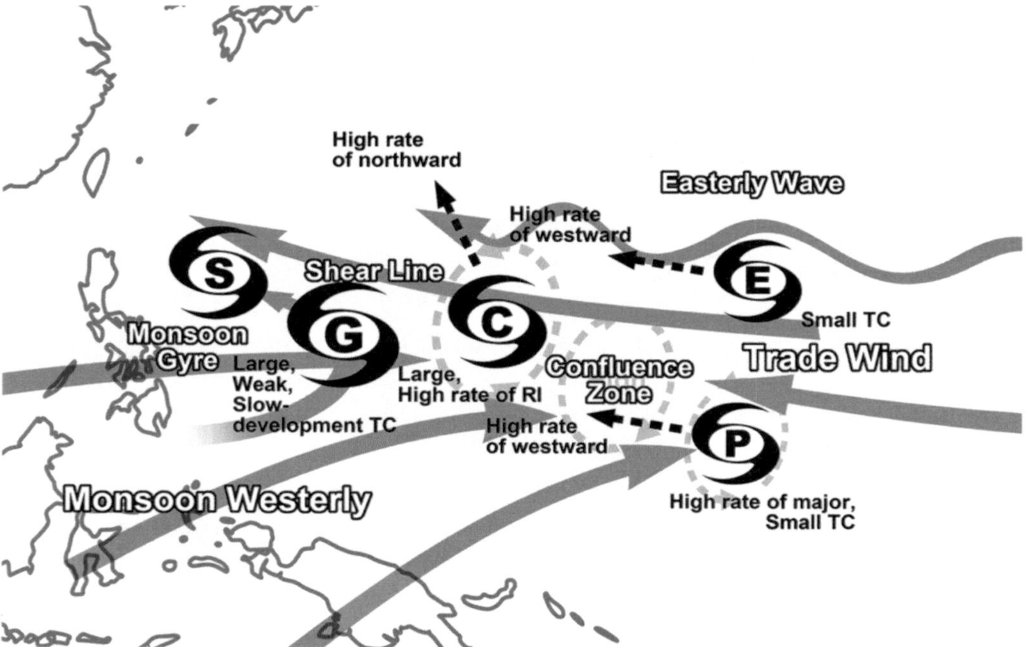

Figure 2.4: Schematic image of the five different flow patterns leading to TCG. TCG location is indicated and the westerly and easterly winds are illustrated by arrows (Fudeyasu and Yoshida, 2018).

2.1.3 Regional differences within the WNP

The favorable synoptic genesis parameters, dynamic and thermodynamic environmental conditions, and thresholds depend on the ocean basin. Therefore, an adequate study of the considered basin is crucial. The SCS is the largest semi-enclosed marginal sea in the WNP, with an area of $3.5 \cdot 10^6\,\mathrm{km}^2$. According to Liang (1991), the East Asian monsoon strongly influences the basin. Southwesterly summer winds occur in May in the southern and central parts and in June in the entire basin. In the northern section, the winter monsoon begins in September and expands over the whole ocean basin in the SCS up to November. In between, there is an area with confluent winds: westerly in the south and easterly in the north. This confluence zone leads to powerful vertical motions. Therefore, the zone is a preferred location for TCG (Sec.2.1.2). Consequently, TCs occur frequently in winter in the southern basin and in summer in the northern part. Environmental factors also influence TC frequency. In winter, the sea surface temperatures are under the threshold of 26° C over the northern and western parts of the SCS (Fig.2.5). Additionally, the vertical wind shear values and relative humidity are unfavorable for TC genesis (Wang et al., 2007, their Fig.2.7, Fig.2.6). In contrast, a reverse distribution of vertical wind shear is seen in summer, with the highest values over $10\,\mathrm{m\,s^{-1}}$ in the southern parts and favorable conditions in the central and northern parts. In winter, the relative vorticity is enhanced close to the coast of the largest island of the Philippines, Luzon , which indicates a preferred location for TCG.

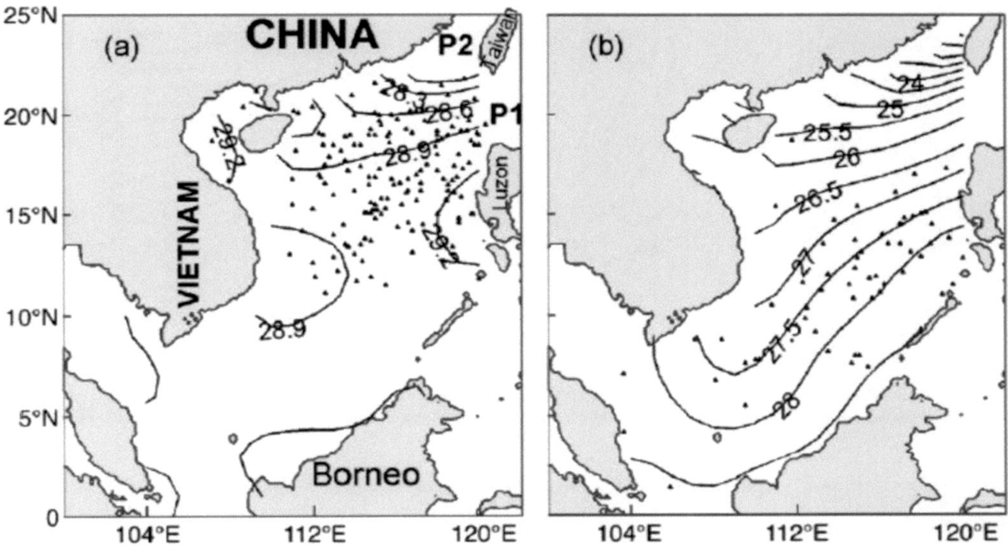

Figure 2.5: Mean climatologic sea surface temperature over the SCS between 1948 and 2003 for southwesterly monsoon (left) and northeasterly monsoon (right). Additionally, the genesis initial points of all storms are indicated (Wang et al., 2007)

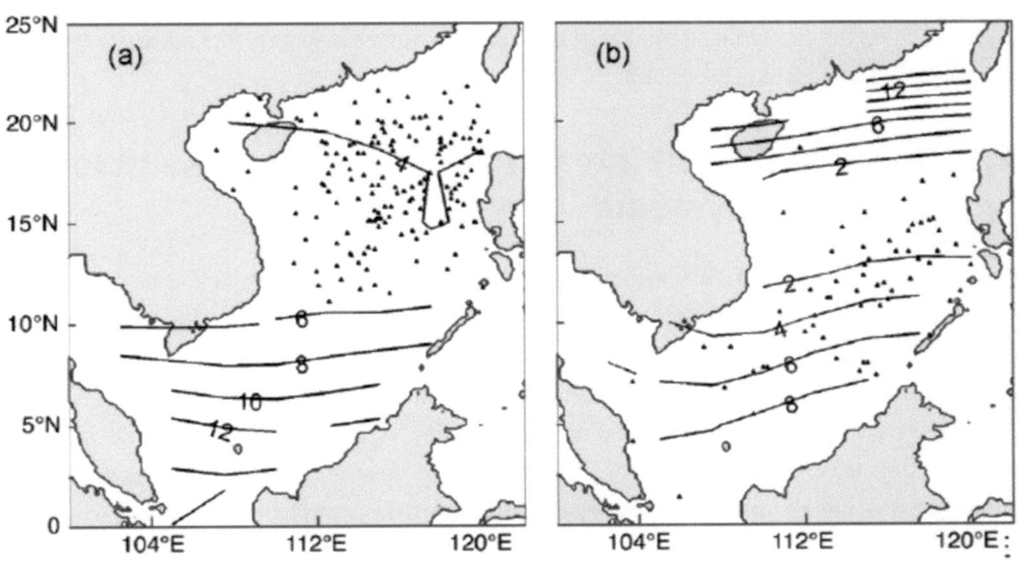

Figure 2.6: Same as (Fig.2.5), but for the vertical wind shear (Wang et al., 2007)

Yuan et al. (2015) investigated the frequency of precursor synoptic-scale disturbances associated with TCG over the SCS between 2000 and 2011. All 35 genesis events occurred between May and December. The authors also observed that most cyclones developed in association with a

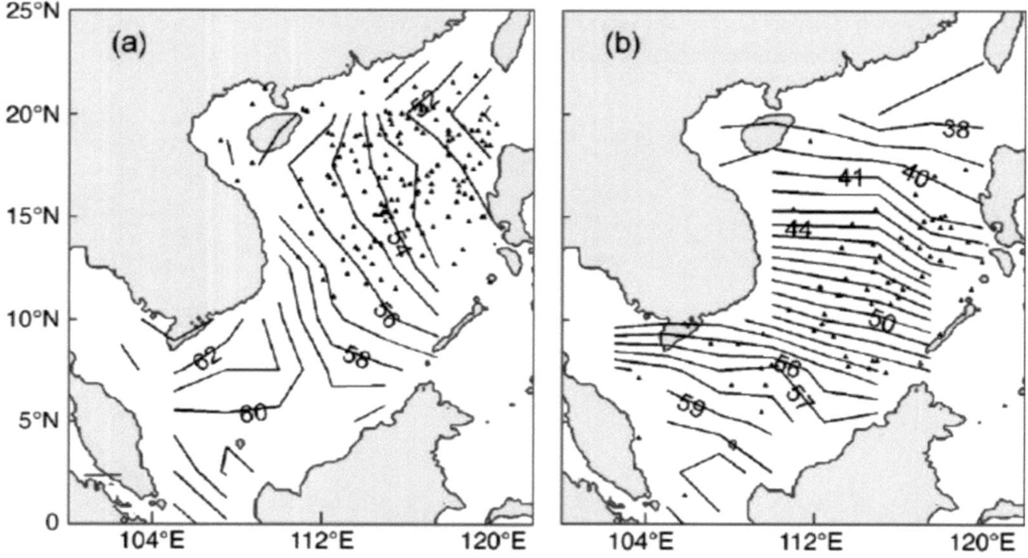

Figure 2.7: Same as Fig.2.5, but for the relative humidity (Wang et al., 2007)

synoptic wave train at the confluence zone followed by the monsoon shear line. This is in contrast to the other parts of the WNP, where the monsoon shear line is the most frequent genesis factor. The WNP is the only ocean basin where TCs develop in all months of the year (McBride, 1995). However, similar to the SCS are the most TCs occurring between May and December (Fig.2.8). In contrast to the SCS, the highest storm frequency is visible earlier in August. This shift and the higher relative frequency in autumn are related to northerly cold surges; that influence on model predicted TCI is studied in (Sec.5.1.4).

2.2 Uncertainties of intensity forecasts in ensemble prediction systems

This study uses perturbed EPS forecasts at ECMWF to identify potential causes of OFC and UFC of TCI. Numerical weather prediction systems are non-linear dynamical systems whose future state depends strongly on the initial conditions and whose estimation of the current state is inaccurate. Additionally, the numerical models have inadequacies due to errors in atmospheric dynamics. These uncertainties contribute to forecasting errors that increase in line with forecast lead time (Leutbecher and Palmer, 2008). The EPS quantifies these flow depending errors. Therefore, the initial conditions are perturbed through data assimilation techniques to account for this insecurity. Additionally, special Kalman filters are used to correct deterministic forecasts and estimate probabilities via observed error distributions (Palmer et al., 2007). Consequently, the EPS includes 50 perturbed and one unperturbed forecast (the high-resolution deterministic run).

Tao and Zhang (2014) have shown the limited predictability of random, chaotic moist convection in the case of slight differences in the initial conditions from the natural state. Additionally, it is stated in this study that these limitations are more apparent under moderate and strong vertical

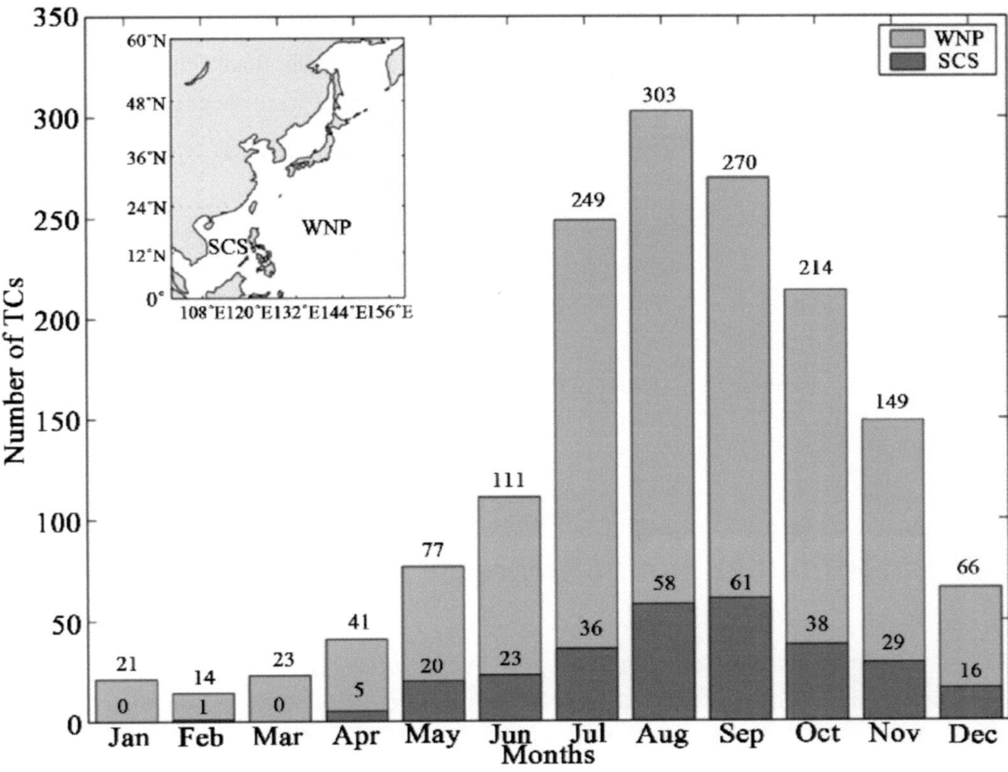

Figure 2.8: Accumulated monthly TC events in the WNP (shaded light grey) and in the SCS (shaded dark grey) during the period 1945-2010 (Xu et al., 2019).

wind shear conditions. This result shows the strong connection between vertical wind shear and moisture in the lower and middle tropospheres. Van Sang et al. (2008) note that small moisture perturbations in the lower troposphere change the flow asymmetry and modify the structure of the deep convective vortex.

Zhang and Tao (2013) investigated the predictability of sheared tropical cyclones explored through a series of convection-permitting ensemble simulations provided by the weather research and forecasting model with different environmental vertical wind shear, sea surface temperatures, and ambient moisture contents. As a result, it is indicated that environmental vertical wind shear significantly affects the intrinsic predictability of TC. This effect is most substantial during the formation and rapid intensification stage. A higher value of vertical wind shear leads to more significant discrepancies in intensity forecasts. These differences are associated with distinct onsets of rapid intensification. It is stated that upscale error growth from differences in moisture leads to contrasting timing of precession and vortex alignment. This process contributes to forecasting uncertainties because a TC intensifies in the forecast model after the tilt, and the shear reaches its minimum.

Emanuel (1986) identified that the main ensemble spread with increasing lead time is amplified by differences in convection-related latent heat fluxes and the wind-induced surface heat exchange

(WISHE) process. Therefore, discrepancies in deep moisture and convective available potential energy have a substantial effect on ensemble evolution. Additionally, Emanuel and Nolan (2004) show the significant bearing of errors in the prediction of vertical wind shear in the vicinity of a TC on the evolution of storm intensity.

In sum, previous studies have indicated that the limited predictability of TCI is possibly related to inaccuracies in the combination of vertical wind shear with moisture in the low- and mid-levels of the troposphere. Furthermore, intensity forecast capability is sensitive to convective activity, which is forced by the vertical flux of mass and heat. In the following section, the concept of low-entropy air ventilation is introduced to examine the interaction of these atmospheric state variables more closely.

2.3 Potential causes of errors in the prediction of TCI

This section describes the assumptions as to why TCs are often falsely predicted. The underlying theoretical processes are analyzed, and different proxies (such as the ventilation index and the modified cold surge index) are explained. Finally, the effect of convective heating on TC intensification is elucidated. Subsequently, the PTE is introduced in the context of the diabatic heating term and its influence on TCI.

2.3.1 Ventilation of low entropy air

As discussed in (Sec.2.1.1) and (Sec.2.2), environmental vertical wind shear plays a crucial role in TCI. Strong vertical wind shear has a detrimental effect on TCI. Various hypotheses have been proposed for this effect. Frank and Ritchie (2001) examined the impact of vertical wind shear in a large-scale environment through a series of numerical simulations. They hypothesized that potential vorticity becomes concentrated in the saturated portions of the eyewall rather than the inner eye, thus eroding TC from the top down. Conversely, Wong and Chan (2004) assume that the reversal of the secondary circulation, which causes entrainment of environmental air at upper levels, is the main contributor to the weakening of TCs due to enhanced vertical wind shear. Another theory is the decrease of TC heat engines by ventilating cyclones with low-entropy air in the mid-levels. As a consequence, convective downdrafts caused by evaporative cooling intrude the low-entropy air in the boundary layer, where it is advected inward by the radial flow (Riemer et al., 2010). Tang and Emanuel (2010) introduced a framework to study the weakening effect of ventilation on TC intensity, which defines a ventilation index to provide evidence for the ventilation process Tang and Emanuel (2012):

$$\Lambda = \frac{U_{\text{shear}} X_{\text{m}}}{U_{\text{pi}}}. \tag{2.5}$$

Here, U_{shear} describes the bulk environmental vertical wind shear between 200 and 850 hPa, X_m is the nondimensional entropy deficit, and U_{pi} is the maximum potential intensity. The entropy deficit is defined as:

$$X_m = \frac{s^*_m \;\; s_m}{s^*_{sst} \;\; s_b}. \tag{2.6}$$

where s^*_m is the saturation entropy in the inner core of the TC at 600 hPa, and s_m is the environmental entropy outside the cyclone. The denominator is the entropy deficit and quantifies the intrusion of low-entropy air in the midlevel into the storm. The nominator measures the strength of the enthalpy fluxes from the ocean surface that are forced against the entropy deficit. s^*_{SST} is the saturation entropy at the sea surface temperature, while s_b is the entropy in the boundary layer. The entropy deficit is calculated at the radius of the maximum winds at maximal TC intensity.

For the computation of entropy, the formula according to Bryan (2008) is used:

$$s = c_p \log(T) - R_d \log(p_d) + \frac{L_{vo} r_v}{T} - r_v R_v \log(H), \tag{2.7}$$

where c_p is the specific heat at constant pressure, R_d is the gas constant for dry air, p_d is the partial pressure for dry air, L_{vo} is the latent heat for vaporization, r_v is the water vapor mixing ratio, R_v is the gas constant for water vapor, and H is the relative humidity. L_{vo} is set at $2.556 \cdot 10^6 \, \mathrm{J\,kg^{-1}}$ to compensate for neglecting water vapor in the entropy.

A theoretical upper bound on maximum possible wind speed is considered by the denominator term in (Eq.2.5). This term combines a local thermodynamic profile and gradient wind balance, according to Bister and Emanuel (2002), and is given by:

$$V_m^2 = c_p(T_s - T_0)\frac{T_s}{T_0}\frac{C_k}{CD}\ln(\Theta_e^*) - \ln(\Theta_e)|_m. \tag{2.8}$$

where V_m is the maximum gradient wind speed, T_s is the sea surface temperature, T_0 is the mean outflow temperature, c_k is the exchange coefficient for enthalpy, C_D is the drag coefficient, Θ_e^* is the saturation equivalent potential temperature at the ocean surface, and Θ_e is the equivalent potential temperature in the boundary layer. The last factor is evaluated at the radius of maximum winds. This relationship between the bulk environmental vertical wind shear, the entropy deficit (Eq.2.6), and the maximum potential intensity (Eq.2.8) describes Tang and Emanuel (2010) TC ventilation framework. A robust amplified flow regime with significant vertical wind speed differences with height tilts the TC and induces structural asymmetries. Suppose the environment indicates a much lower entropy relative to the TC; the more significant the entropy deficit, the more low-entropy air intrudes into the high entropy reservoir of the low-level TC center (Fig.2.9). Intrusion reduces the energy supply by surface fluxes. As a result, the available potential energy decreases, and the conversion to mechanical energy is limited. More substantial surface fluxes (indicated by a higher potential intensity) can reduce the strength of the intrusion of low-entropy air. If the entropy deficit or the shear term is zero, the ventilation vanishes, and the TC reaches its maximum potential intensity. Increasing the ventilation index enhances the shear term and midlevel entropy concerning the surface fluxes. Consequently, the TC intensity decreases below its maximum potential intensity.

If the ventilation overcompensates the surface fluxes, the TC decays.

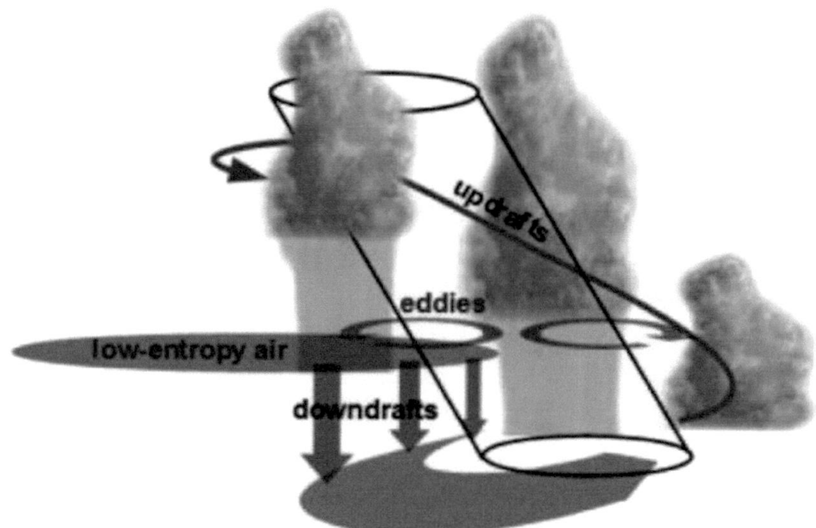

Figure 2.9: Illustration of the Ventilation process into a TC in an enhanced environmentally sheared flow. Tilting the vortex cause convective asymmetries, which both excite mesoscale eddies. These eddies transport low entropy air from the midlevel into the TC center and weaken the helical updrafts. Evaporational cooling induces downdrafts, which are flushed into the high- entropy air and counteract the generation of available potential energy by surface fluxes and reducing the TC intensity (Tang and Emanuel, 2012)

Tang and Emanuel (2012) calculated the seasonally averaged ventilation index between 1990 and 2009 from interim ECMWF reanalysis data. This index is located in the tropics, where TCs frequently develop under 0.1 (Fig.2.10). Moreover, the ventilation indices over all seasons are under 0.1. This threshold defines areas with high maximum potential intensities over $80\,\mathrm{m\,s^{-1}}$, relative low vertical wind shear under $15\,\mathrm{m\,s^{-1}}$, and low entropy deficits below 0.7.

2.3.2 Northerly cold surges

Northerly cold surges over the SCS frequently occur during the East Asian winter monsoon between October and March (Chan and Li, 2004). They are classified as pressure surges with strong north winds, anomalously low temperatures, and an increase in pressure (Compo et al., 1999). According to Boyle (1987), cold surges are defined by arbitrary thresholds of 12- to 24-hour temperature change, wind acceleration, wind speed, surface pressure change, and station-based differences. Compo et al. (1999) investigated the linear statistical relationship between pressure surges, tropical convection, and tropospheric circulation using a 10-year dataset (1985–1994) of ECMWF operational analyses. Their findings indicate relationships between sub-monthly surges over the SCS and tropical circulation. Thus, enhanced convective activity to the south of Indonesia and over the SCS, the east Indian Ocean, and the Philippines are observed. Additionally, sub-monthly cold surges over the Philippine Sea are associated with westerly wind activity and strengthened convective activity over the WNP. As a result, cold surges are able to promote the

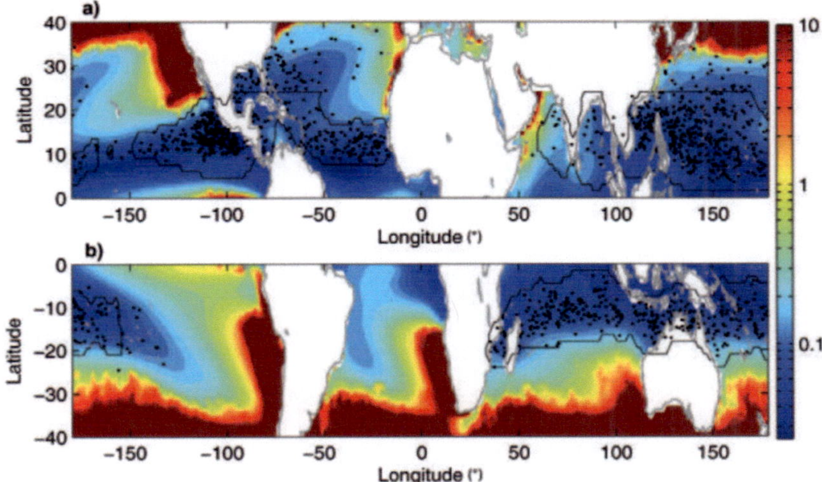

Figure 2.10: a) July-October ventilation index for the northern hemisphere; b) December-March ventilation index for the southern hemisphere, both averaged over 1990-2009. Black dots indicate TCG events over the period (Tang and Emanuel, 2012)

development of tropical cyclones. Chang et al. (2003) indicates that tropical storms formed over the maritime continent are often related to strong northerly cold surges over the SCS. Additionally, pressure surges enhance the convergence in a TC. If a pressure surge flows outside along a TC, the increased pressure difference between the outer area of the cyclone and the storm center fosters the inward flow of water vapor and, therefore, the externally forced convergnece (Gray, 1998). Consequently, the intensity increases if all other conditions remain constant.

To compute an index that describes a northerly cold surge event in the SCS, Yokoi et al. (2009) used the three-day mean meridional wind anomaly at 925 hPa averaged over $110-120°$ E along $20°$ N. As a proxy for the temperature drop, the three-day mean temperature anomalies at 850 hPa averaged over $105-115°$ E along $25°$ N are calculated. If the mean meridional wind anomaly is a temporal minimum and larger than $3.1\,\mathrm{m\,s^{-1}}$, and the temperature anomaly has a temporal minimum in the five to two days before to two days after the meridional wind minimum, the event is characterized as a cold surge. In contrast, Chang et al. (2005) only considered the mean meridional wind at 925 hPa between $110-117{,}5°$ E along $15°$ N. The averaged mean meridional wind is used to describe the strength of the cold surge event: values under $8\,\mathrm{m\,s^{-1}}$ indicate no event, between 8 and $10\,\mathrm{m\,s^{-1}}$ a weak event, between 10 and $12\,\mathrm{m\,s^{-1}}$ a moderate event, and over $12\,\mathrm{m\,s^{-1}}$ a strong event.

In this study, a modified cold surge index is calculated using the primary aspects according to (Chang et al., 2005). The index is computed on a storm-relative basis to estimate the maximum influence of the surge on the TC (Sec.4.4).

2.3.3 Water vapor fluxes

The inward-directed and the locally generated are two types of water vapor fluxes that play a crucial role in TCI. The former one is forced by the secondary circulation and is therefore strongly dependent on TC intensity (Sec.2.1). According to Gray (1998), the flux of water vapor converges in the inner regions of a cyclone and intensifies convective activity. This moisture convergence is triggered by a strong wind surge around the cyclone, through differential heating between the cloudy and the cloud-free areas in the higher troposphere or by frictionally and internally forced convergence. This moisture flux is defined by two parts of the water budget equation Wu et al. (2013):

$$C = -\frac{1}{g} = \int_{p_\mathrm{T}}^{p_\mathrm{s}} (q\nabla \cdot \boldsymbol{V})\mathrm{d}p - \frac{1}{g} \int_{p_\mathrm{T}}^{p_\mathrm{s}} \boldsymbol{V} \cdot \nabla q \mathrm{d}p. \tag{2.9}$$

where q is the specific humidity, **V** is the wind vector, g is the acceleration gravity, p_T defines the pressure at the top of the atmosphere, and p_S the surface pressure. The first term describes wind convergence, while the second gives the moisture advection. Takakura et al. (2018) investigated the water origins in the vicinity of a mature TC and found that a significant amount of total precipitable water around the cyclone comes from external regions rather than the local sea surface. This result is consistent with previous studies (Braun (2006); Yang et al. (2011); Fritz and Wang (2014)). These studies evaluated water budgets in the inner core of a TC and found that the water vapor transport from the sea surface to the inner core area only accounted for a small proportion of the horizontal water vapor transport. Additionally, the net horizontal water vapor transport accounted for a substantial part of the net condensation in the inner area of the storm. Through a high-resolution numerical model simulation of Typhoon Nari, Yang et al. (2011) show the importance of moisture advection (from the surroundings beyond the outer rainbands into the storm center) in enhancing an existing TC. To examine these processes, both terms of (Eq.2.9) are evaluated in this work at different layers in the boundary layer from the surface to 850 hPa. Additionally, to consider the vertically integrated water vapor flux, both terms of (Eq.2.9) are integrated from the surface to the upper troposphere. This accounts for the majority of the total condensation in the inner core of a TC (Fritz and Wang, 2014). The vertical flux is also enhanced by the strengthening of low-level convergence associated with secondary circulation. As a result, the amount of vertically integrated water vapor flux in the TC area depends on the storm's intensity.

2.3.4 Convective heating

TCs are rotating systems with a circular, relatively small band of organized deep convection around their storm centers. Latent heat release, driven by the condensation of water vapor, fosters secondary circulation, which is responsible for the transport of water vapor into the TC center from the evaporation of warm waters in the boundary layer (McBride, 1995). According to (Raymond and López Carrillo (2011); Fritz and Wang (2014)), positive feedback is seen between enhanced moisture convergence, higher precipitation rates, and more intense convective activity in a TC, which reverses the negative feedback of convection on water vapor and makes convection in a developing TC more sustainable. As a result, local evaporation decreases, and the vertically integrated inward-directed water vapor flux increases, especially after the TCG due to the deepening inflow layer (Fritz and Wang, 2014). Consequently, cold surges are possible triggers of enhanced convection in a cyclone and thus influence its diabatic heating rates (Sec.2.3.2).

Using idealized numerical simulations, Nolan et al. (2007) found that a nearly saturated inner core contributes to the rapid development of a TC. Specifically, an inner core with relative humidity above 80% over a deep column and a strengthening midlevel vortex leads to a rapidly intensified vortex near the surface. Before genesis, the strengthened midlevel vortex contributes to increased efficiency in converting latent heat energy into kinetic energy in the cyclonic wind field. As a result, the tangential winds increase, and the TC intensifies. Similarly, Smith and Montgomery (2016a) have noted the higher efficiency of diabatic heating inside the radius of maximum winds in a rapidly developing warm-core thermal structure, which results in increases in intensity and tangential winds. Additionally, (Vigh and Schubert, 2009) relate the higher efficiency of diabatic heating in the inner core of a TC to higher inertial stability as the vortex intensifies.

In summary, convective heating is a crucial factor in TCI. The extent of the TCI depends on several factors, such as the environmental thermodynamical and dynamical conditions, the spatial distribution, and the intensity of the TC itself. Therefore, an analysis of the amount of diabatic heating in TCI is applied in this study using the PTE equation (Fink et al., 2012, Sec.4.5).

3 Research Questions

This chapter discusses the main research questions in this work. The predictability of TC intensity is studied through the EPS of the ECMWF model based on three case studies (two over-forecasting events and one under-forecasting event). K-means clustering is applied to quantify the difference in intensity and location of the simulated TCs according to individual perturbed forecasts after 120 h and 84 h forecast time. Furthermore, environmental variables with the most influential intensity forecast sensitivity are analyzed by illustrating calculating indices, including the ventilation index (VI). These proxies are compared through the computation of normalized differences. The study area is the WNP, including the SCS. Only TC Atsani shows movement over the SCS after an increased lead time for a fraction of members in the ensemble. Typhoon Kammuri affected the central and western WNP and made landfall in the Philippines, while Super Typhoon Champi mainly influenced the central WNP.
Thus, the following research questions are defined:

1. **How does the intensity of TCs evolve in the ensemble forecasts for increasing lead times?** The members in the EPS are clustered according to their pressure after different forecast times to analyze their respective intensity tendencies. Time evolution is illustrated through time series of mean intensity and standard deviations of the corresponding clusters. Critical forecast times with growing spreads between the most appropriate clusters are identified.

2. **What is the role of sensitive acting environmental variables on TCI?** Sensitive acting environmental variables like atmospheric moisture or the deep-layer vertical wind shear are examined through spatial storm-relative composites and normalized differences between the considered clusters. The illustration of the VI combines these parameters and describes the influence of dry air intrusions into the TC center.

3. **Is there an influence from long-lasting northerly cold surges?** The assumptions of the overestimated positive effect of these cold surges on TCI are examined for TC Atsani. Different members are assigned to clusters through k-means clustering based on cold surge intensity, described by the meridional wind connected to the cyclone at the top of the boundary layer. Furthermore, the clusters are compared at the critical forecast points with the maximal simulated cold surge strength in the intensification stage.

4. **Do discrepancies in water vapor fluxes contribute to an overestimation or underestimation of TCI?** The inconsistencies in vertically integrated water vapor fluxes are illustrated using time series of normalized differences and storm-relative composites to examine whether water vapor fluxes play a crucial role in the intensification stage of the intensive cluster. Additionally, the terms of the water budget equation are contemplated.

5. **Are there variances in the potential causes between individual case studies?** Contrasting conditions depending on the TC formation and intensification process across different locations lead to discrepancies in potential causes of over- and under-forecasting events. Comparing the most sustainable variables for intensification between different storms resolves this issue.

4 Data and Methods

This chapter describes the employed data structures and sets, followed by the applied data organization and analysis methods. A synoptic description of the case studies is presented, followed by a methodological explanation of the calculation of the storm-relative cold surge index. Finally, the measurements for the various indices and the results from the PTE are outlined.

4.1 Ensemble prediction system

To identify the location of the analyzed TCs, the TIGGE model tropical cyclone track data were used. This dataset includes ensemble-generated TC track data from the ECMWF, the UK Met Office, the US National Centers for Environmental Prediction, Japan Meteorological Agency, China's Meteorological Administration, the Meteorological Service of Canada, MeteoFrance, and Korea's Meteorological Administration (Swinbank et al., 2016). It holds all data from the Observing System Research and Predictability Experiment (THORPEX), which aims to accelerate improvements in the forecasting of high-impact weather events. TIGGE is one of the top programs of THORPEX, providing operational ensemble forecasts. In turn, it is utilized in a range of predictability and dynamical studies (Swinbank et al., 2016).

To determine the location of a TC with a warm core, TIGGE uses a tropical cyclone tracker (Van der Grijn et al., 2005). A storm appears in the dataset if the deterministic forecast and the EPS identify a maximum wind speed of $17\,\mathrm{m\,s^{-1}}$ or more. The time step of 6 h allows a high time resolution, which provides higher recognizability for small-time intensity and location variability. The algorithm can restart the tracking at the same feature up to 24 hours later. These tracking interruptions are crucial aspects due to the transition of small islands, where a TC often shows a short disappear time. The tracker detects any storm that develops during forecasts, even though it may not be present in the initial conditions; the only requirement is that the cyclone must be present for a minimum of two time steps (i.e., 12 h). The data include all members of the EPS. As a result, the individual members are comparable in their track location and their different environmental conditions through the EPS data. In this work, the TIGGE database is utilized to determine the center point of the individual simulated cyclones in the ensemble members. Additionally, their intensity is determined based on their minimum central pressure and maximum wind speed.

The EPS by ECMWF quantifies the uncertainty using a set of 50 forecasts, starting from slightly different initial states (Leutbecher and Palmer, 2008). These states are close but not identical to the best estimates of model equations. The inaccuracies are a result of the uncertainties in the initial conditions and the model. Measurement errors and the heterogeneous spatial distribution lead to imprecision in the initial state. Unavoidable simplifications in representing the complexity of nature in the numerical models also contribute to these initial discrepancies. At the beginning of a forecast, these minor errors can be amplified and result in a significant mistake in increasing forecast time. The divergence of the control run plus the 50 members estimates the error at a particular time. A smaller spread indicates a predictable atmosphere, while a larger one shows the constraints in predictability. In this work, a resolution of 32 km is utilized for the different state variables. The environmental variables are projected on the grid with a distance of $0.125° \times 0.125°$. A grid distance of $0.5° \times 0.5°$ is utilized for the wind field. This high resolution allows for more profound insight into the small-scale processes that contribute to TCI.

4.2 Storm identification and case studies

The tropical cyclone track database (Sec.4.1) is employed to identify storms with a strong OFC and UFC of intensification. Three kinds of cyclones were chosen:

1. Cyclones that appeared only in the members of the EPS and not in the real world.
2. Storms that showed, after a forecast time between 72 and 120 h, a much higher/lower intensity than in real observations.
3. Events with the most intense OFC and UFC from September to November between 2011 and 2020 were selected.

The TIGGE dataset features 97 events in the WNP over the selected period that show a TC with wind speeds over $17\,\mathrm{m\,s^{-1}}$ for at least a forecast time of 120 h (Fig.4.1). The highest number of TCs occurs in August, with 52 events. From September to November, the frequency decreased steadily from 41 to 21 events. This reduction is mainly related to seasonal changes in sea surface temperatures. In September, the area from the equator to southern Japan has sea surface temperatures of 26° C or above on average, and large regions in the middle and south basin demonstrate values above 28° C (McBride, 1995). In comparison, this area is restricted to the southern parts of the WNP in November. Additionally, the baroclinic zone with high vertical wind shear is extended to the south in late autumn. These high values, in combination with the lower sea surface temperatures and the southward-moving monsoon confluence zone, contribute to a decrease in TC activity in November.

The mean error is calculated according to Hamill et al. (2011) to identify these cyclones

$$P(t) = \frac{\sum_{i=1}^{m} P_{\mathrm{i}}(t)}{m}. \tag{4.1}$$

$P(t)$ is the mean error of the pressure after 120 h between the model and the observations (mean intensity error). $P_{\mathrm{i}}(t)$ is the error between the individual ensemble members and the natural state, and m is the number of all members. A negative value represents an OFC event, whereas a positive one defines a UFC event. The evolution with time is also a critical selection contributor: storms with the most substantial increase in negative/positive values of the mean intensity error over time indicate the most intense OFC/UFC events. Therefore, the 12 h mean intensity error change is

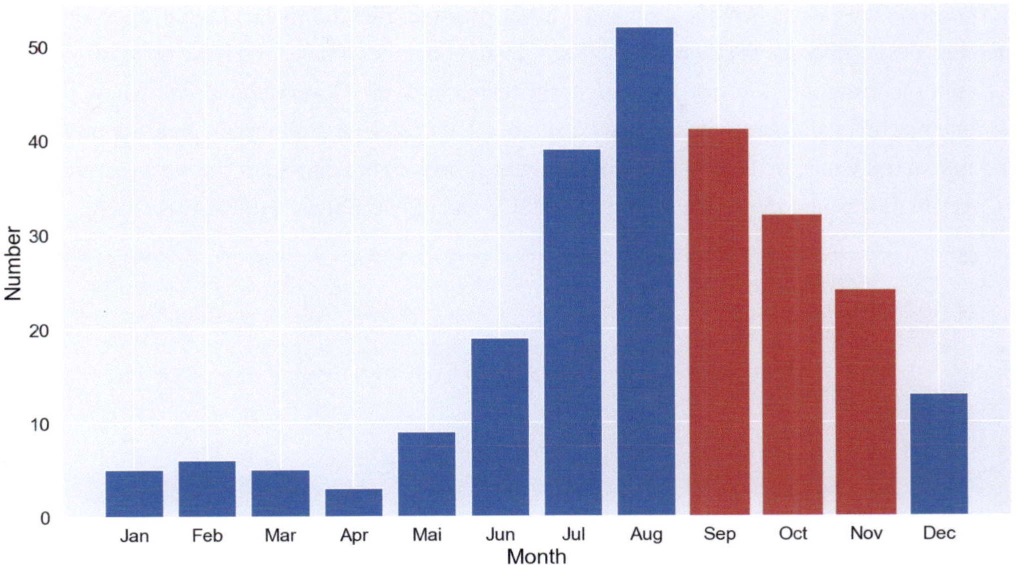

Figure 4.1: Number of tropical storms with winds over $17\,\mathrm{m\,s^{-1}}$ per month over the WNP in the period from 2011 to 2020. Red bars indicate the selected period (between September and November).

evaluated here (Fig.4.2).

The second condition is an extensive range in the ensemble. This reviews a predictable atmosphere with a slight divergence or an unpredictable one with a significant spread. Furthermore, to identify potential causes that lead to OFC or UFC, it is crucial to have considerable variation in the ensemble to compare clusters of ensembles close to the observation with one far away. As a result, a higher spread from the actual value contributes to a more accurate analysis. The mean spread is computed as follows:

$$S(t) = \frac{\sum_{i=1}^{m} S_{\mathrm{i}}(t)}{m}. \tag{4.2}$$

where $S_{\mathrm{i}}(t)$ is the absolute error between the individual ensemble members and the ensemble mean. The storms with the highest mean absolute errors and the most robust spread are analyzed here.

The first case study event is TC Atsani, which formed over the WNP on October 29, 2020, in the wake of the preexisting Super Typhoon Goni. The storm growed due to Rossby wave energy dispersion and favorable environmental conditions (Sec.2.1.2). As Goni moves with the mean flow against the strongest planetary vorticity gradient to the west, Rossby waves emit energy eastward. Consequently, a Rossby wave train with alternating anticyclonic and cyclonic vorticity perturbations forms to the east; TC Atsani develops from one of the cyclonic vortices (Fig.4.3). On November 1, 2020, the ECMWF ensemble forecasting model simulated that the cyclone would move to the northwest and then, after 36 to 60 h forecast time (dependent on the corresponding member), make an anticyclonic loop close to the northeastern coast of Luzon. After this loop, the members in the ensemble indicated strong deviations in track and intensity (Fig.4.4). The observed storm shows a weaker intensification and a lower pressure after 120 h as a high proportion of the members, especially the southern ones. (Fig.4.2) demonstrates a mean intensity error of −7.8 hPa between the ensemble forecast and the storm that occurred. Observational data show that the cyclone reached its highest intensity on November 4 at 06 Z with a minimum pressure of 993 hPa and maximum winds of 100 km h^{-1} 24 hours later. Forty-eight hours after the strongest intensity, TC Atsani moves into the northern part of the SCS and affects Taiwan (IBTRACS, 2020).

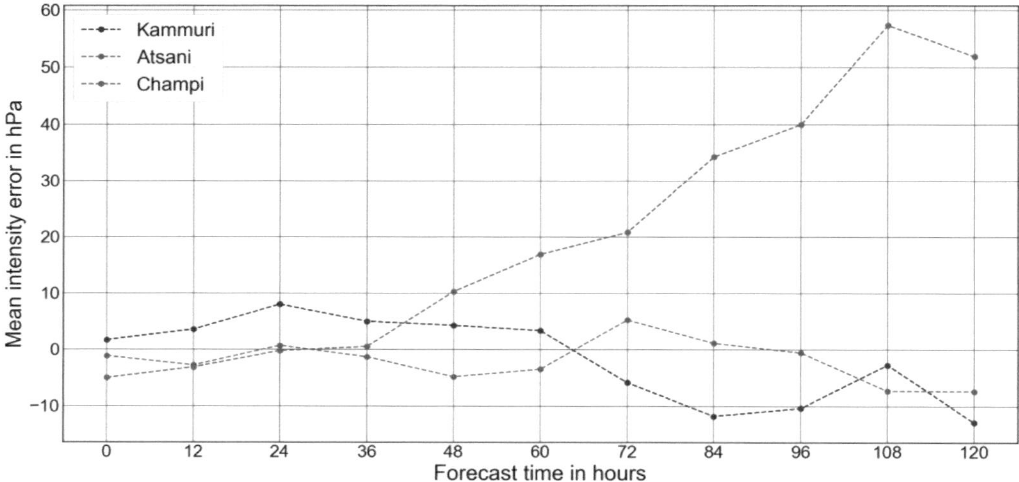

Figure 4.2: Mean intensity error evolution in comparison to observations (IBTRACS, 2020, 2019, 2015), for TC Kammuri (blue), Atsani (green), and Champi (red). Negative values demonstrate an OFC event, positive values an UFC event.

The second storm is TC Kammuri. Kammuri evolved on November 24, 2019, from a convective system influenced by an easterly wave 100 km south of the Mariana Islands in the central WNP. The cyclone moves to the west with only weak intensity changes. This small diversity was mainly due to a strong deep-layer vertical wind shear with values above 15 m s^{-1}. On November 28, 2019, at the beginning of the forecast period, the ensemble forecast indicates a movement toward the north. The storm tracks then display a cyclonal turning toward the Philippines. Subsequently, however, the predicted intensities in the ensemble differ strongly. A substantial proportion of the ensemble, and the high-resolution deterministic run in particular, show a more intense decrease in pressure compared to the observation. The mean intensity error decreased up to 84 h forecast time for values under −10 h (Fig.4.2). At the same time, the spread examines the largest extension (not shown). Consequently, this forecast time is the most appropriate for the k-mean clustering (Sec.4.3.1). On December 2 at 12 Z, the typhoon showed its peak intensity with a minimum pressure of 950 hPa and wind speeds of 61 m s^{-1}, before the storm hit the Philippines and weakened drastically (IBTRACS, 2019).

TC Champi is the case study with the strongest under-forecasting in the dataset: the mean intensity error exceeded 108 h, 50 hPa (Fig.4.2). These discrepancies between the individual ensemble members and the observation demonstrate the inability of the model to simulate the atmospheric conditions relevant to TCI in an adequate semantic way. Champi developed on October 13, 2015, in the central WNP, close to Micronesia. The cyclone subsequently moved west under weak

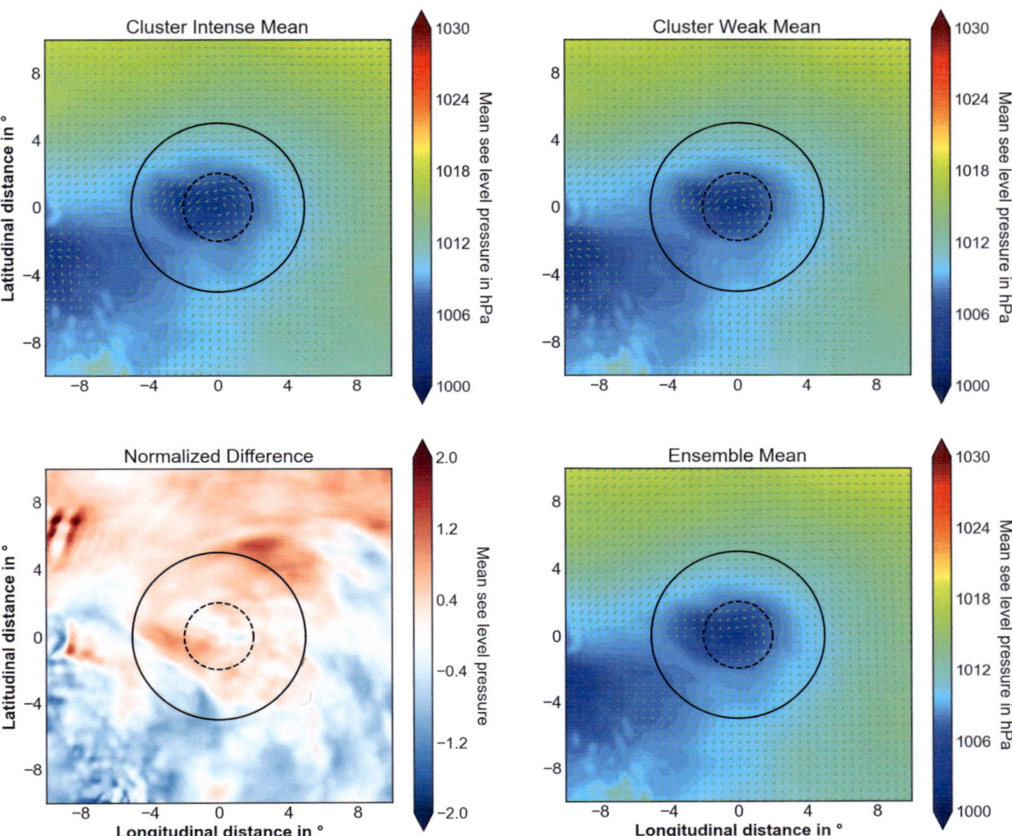

Figure 4.3: Storm-relative composites of MSLP for TC Atsani at initial time on November 01 at 12 UTC, for weak cluster weak (top-left), intense cluster intense (top-right), normalized diffference (bottom-left), and ensemble mean (bottom-right). Typhoon Goni is located to the west, outside of the detail graph.

intensification (IBTRACS, 2015). After October 17 at 00 UTC (i.e., after 72 h forecast time), the track indicates a northwest translation and intensifies rapidly. At this point, the model spread increases, but no member predicts the intensification rates close to the observation. The maximum intensity was observed on October 18 at 12Z with a minimum central pressure of 929 hPa and a maximum wind speed of 64 $m\,s^{-1}$. Champi then weakened continuously due to lower sea surface temperatures and a detrimental deep-layer vertical wind shear and made an extratropical transition on October 26 at 18Z in the central part of the northern WNP.

4.3 Data organization and analysis

In this study, the k-means clustering method organizes the data into sets with similar properties. The data are clustered according to intensity and location as described by the latitudinal and longitudinal positions after 120 h or 84 h. The clustering time is the time with the first maximum/minimum in the mean intensity error with increasing lead time. A proxy for the intensity is the minimum pressure in

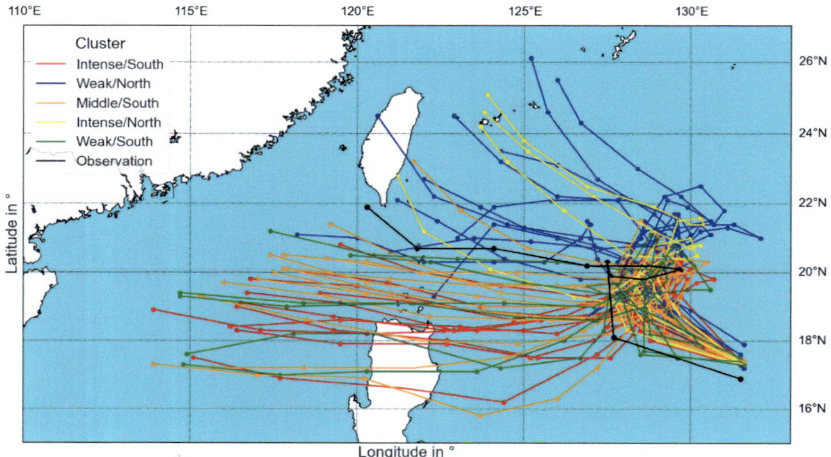

Figure 4.4: Tracks of TC Atsani for the 45 members who are showing a TC after 120h. Different clusters are dependent on TC intensity and location. The black dashed line shows the observed track (IBTRACS, 2020). Time steps are in 12h intervals, beginning at the initial time on November 01, 2020.

the center of the TC. The clusters with the slightest difference in location and the strongest intensity distinctions are then compared. To compare the environmental variables between different vertical levels, the normalized differences of storm-relative composites are computed. A storm-relative approach allows the comparison of properties between two or more storms at various locations. This consideration is used in the calculation of the cold surge index (Sec.4.4). Finally, the computation of the PTE is described.

4.3.1 K-means clustering

K-means clustering is an algorithm based on a centroid-based method whereby different observations were assigned to k-clusters (Hartigan and Wong, 1979). Therefore, a prescribed number of squares within a cluster is minimized. The algorithm requires randomly selected k initial cluster centers. The Euclidean distance between the data points assigns the individual input elements to their nearest cluster center. Subsequently, the allocated components of a cluster are averaged to define a new cluster center; these steps are then repeated until none of the cluster assignments change. Before the algorithm runs, determining the optimum number of clusters is crucial. Two methods are applied in this work: the graphical interpretation using the elbow method and the objective method using the calculation of the silhouette coefficient (Kumpf et al., 2018). For the elbow method, the number of clusters is visualized to the sum of their squared distances, which describes the difference between each point and the centroid in a collection (Fig.4.6). The main target here is to decrease the number of clusters and the sum of squared distances. Thus, the best number of clusters is the point at which the curve shows a knee (or elbow). An overly high number contributes to overfitting, while a lower number leads to underfitting.

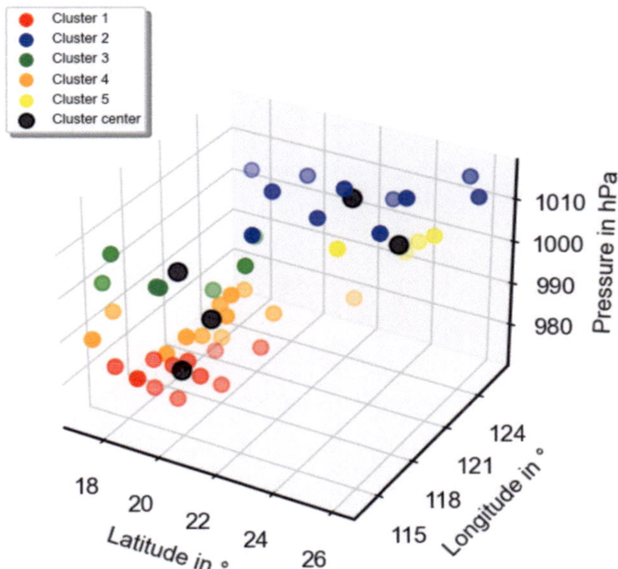

Figure 4.5: 3D Point cloud after k-mean clustering for TC Atsani. Scatter points indicate, the longitude, latitude, and pressure of the individual members for the five clusters. The cluster center is illustrated by the large black scatter dot and describes the average of all points (elements) that belong to the corresponding cluster. The blue cluster possesses the highest pressure mean, while the red one has the lowest (after 120 h forecast time).

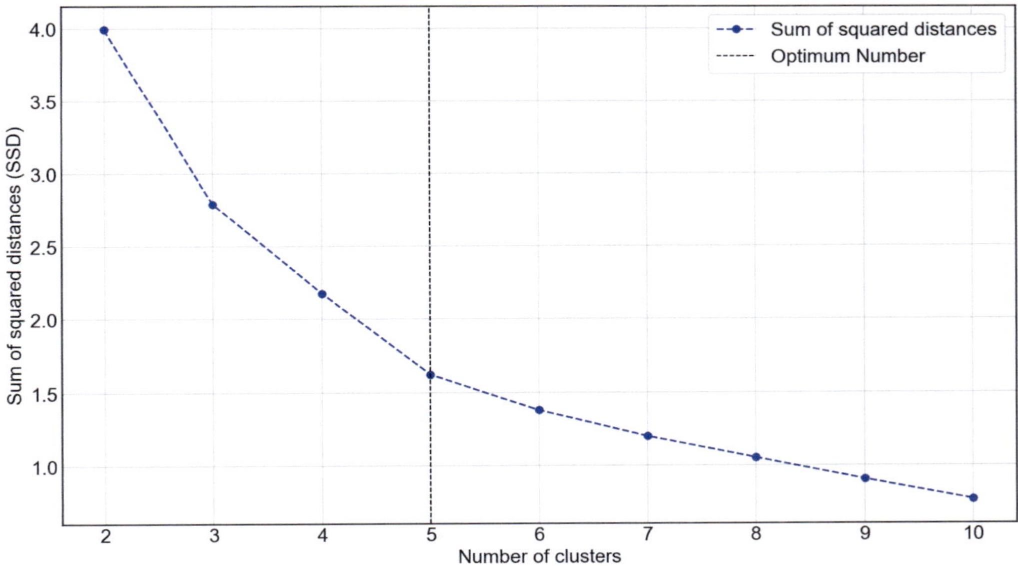

Figure 4.6: Graphical interpretation off the elbow method for TC Atsani. The black dashed line show the optimum number of clusters.

The second method is defined as follows:

$$s(i) = \frac{b(i) - a(i)}{max(a(i), b(i))}. \tag{4.3}$$

$a(i)$ is the mean intra-cluster distance and the mean $b(i)$ nearest cluster distance. A value of 1 indicates that the data points represent the centroid, while a value of 0 shows that the data object does not fit any created clusters, as the distance to any cluster center is the same.

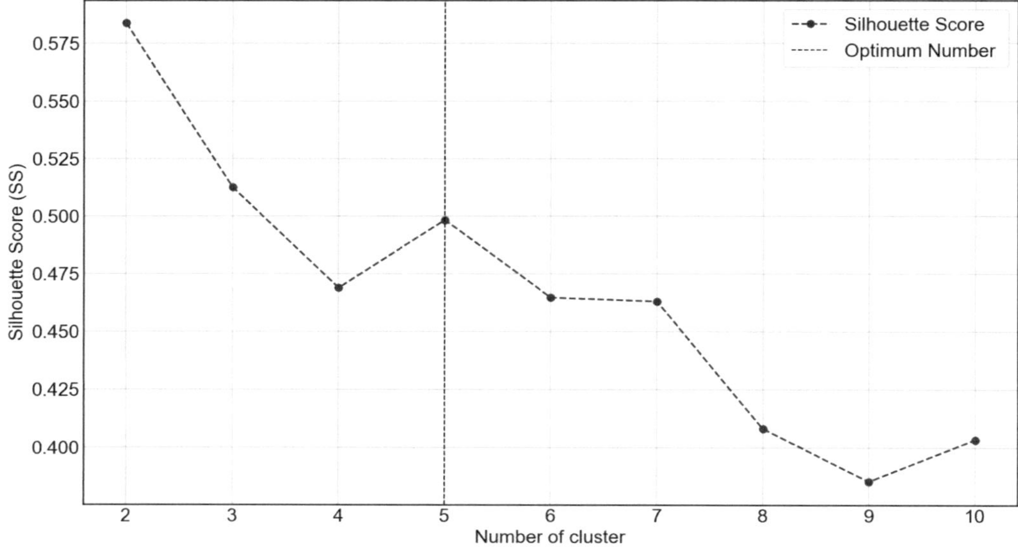

Figure 4.7: Same as Fig.4.6 but for the silhouette coefficient.

In turn, the silhouette coefficient plot displays how close each point in one cluster is to points in the neighboring clusters(Fig.4.7). This method provides a way to assess parameters like the number of groups visually.

Table 4.1: Centers and number of storms for each cluster for TC Atsani after 120 h, Typhoon Kammuri after 84 h and Super Typhoon Champi after 120 h forecast time. Bold text highlights clusters used in this study.

Storm	Cluster	Latitude [°N]	Longitude [°E]	Pressure [hPa]	Number
Atsani	**Intense/South**	**19.15**	**116.74**	**979.18**	**11**
	Weak/North	22.8	122.42	1013.09	11
	Weak/South	**19.21**	**116.05**	**1004.83**	**6**
	Middle/South	19.96	117.40	991.76	13
	Intense/North	24.17	123.15	1003	4
Kammuri	**Intense 1**	**14.84**	**125.42**	**938.00**	**10**
	Intense 2	14.48	124.57	951.61	21
	Weak 2	**14.83**	**125.58**	**962.78**	**14**
	Weak	14.18	123.64	994	5
Champi	**Intense**	**22.81**	**141.91**	**965.62**	**8**
	Middle	21.98	140.20	984.95	21
	Weak	**20.75**	**139.78**	**999.38**	**21**

Both methods are considered to choose the optimum number of clusters, which are then divided into different categories according to their mean intensity and location at 120 and 84 h (Tab.4.1). Location discrepancies must be as minor as possible. Therefore, the cluster with the highest intensity is contrasted with the lowest, while the latitudinal and longitudinal distances are kept as small as possible. Additionally, the weak cluster for the over-forecasting and the strong cluster for the under-forecasting are the closest to the observational intensity at the corresponding forecast time. Consequently, the Intense/South cluster is compared with the Weak/South cluster for TC Atsani. The weakest cluster, in this case, possesses too wide a discrepancy in the location for an adequate analysis. A latitudinal distance of more than 3° contributes to significant differences in environmental forcing. Therefore, the slight differences in location between the southern clusters provide a more accurate analysis of the processes, contributing to discrepancies in the storm's intensity after 120 h.

4.3.2 Storm-relative composites and normalized differences

A storm-relative contemplation is applied to analyze the environmental atmospheric state variables. For every member in the model, the storm center is the center of the composite. The mean of the different clusters shows their environmental conditions. This method has a considerable advantage in that storms with distinct locations are thus comparable. Additionally, to gain insight into the model-predicted inner core processes of a TC, the mean state variable of every cluster is calculated in an area circle with a radius depending on their most significant impact on intensity changes. The center is in the middle of the storm, determined from the tropical cyclone track data (Sec.4.1). The normalized difference is calculated from the different storm-relative fields according to Torn et al. (2015), defined as follows:

$$\Delta x_{\mathrm{i}} = \frac{x_{\mathrm{i}}^{\mathrm{weak}} - x_{\mathrm{i}}^{\mathrm{intense}}}{\sigma_{x_i}}. \tag{4.4}$$

where $x_{\mathrm{i}}^{\mathrm{weak}}$ is the mean value of the weak cluster, which is determined over the k-mean clustering (Sec.4.3.1). $x_{\mathrm{i}}^{\mathrm{intense}}$ is the mean value of the intense cluster and the standard deviation over all members. This normalized difference or standardized anomaly ensures comparison between different vertical levels, fields, and times. This computation is appropriate for analyzing differences at one vertical level since it considers large gradients, providing large absolute differences due to displacement errors.

4.4 Storm-relative cold surge index

The cold surge index is computed according to (Chan, 2005). In that study, the averaged 925 hPa meridional wind over the SCS between 110° and 117,5° E along 15° (Sec.2.3.2) is considered. This method is applied because the simulated cyclones in the distinct members are not all located over the SCS. Here, the cold air masses are advected over a long distance over the open sea. As a result, the temperature criterion according to Yokoi et al. (2009, Sec.2.3.2) is not appropriate because the warm ocean water reduces the temperature and increases the humidity of the air masses. Additionally, the cyclone moves over a period, and its circulation influences the index if

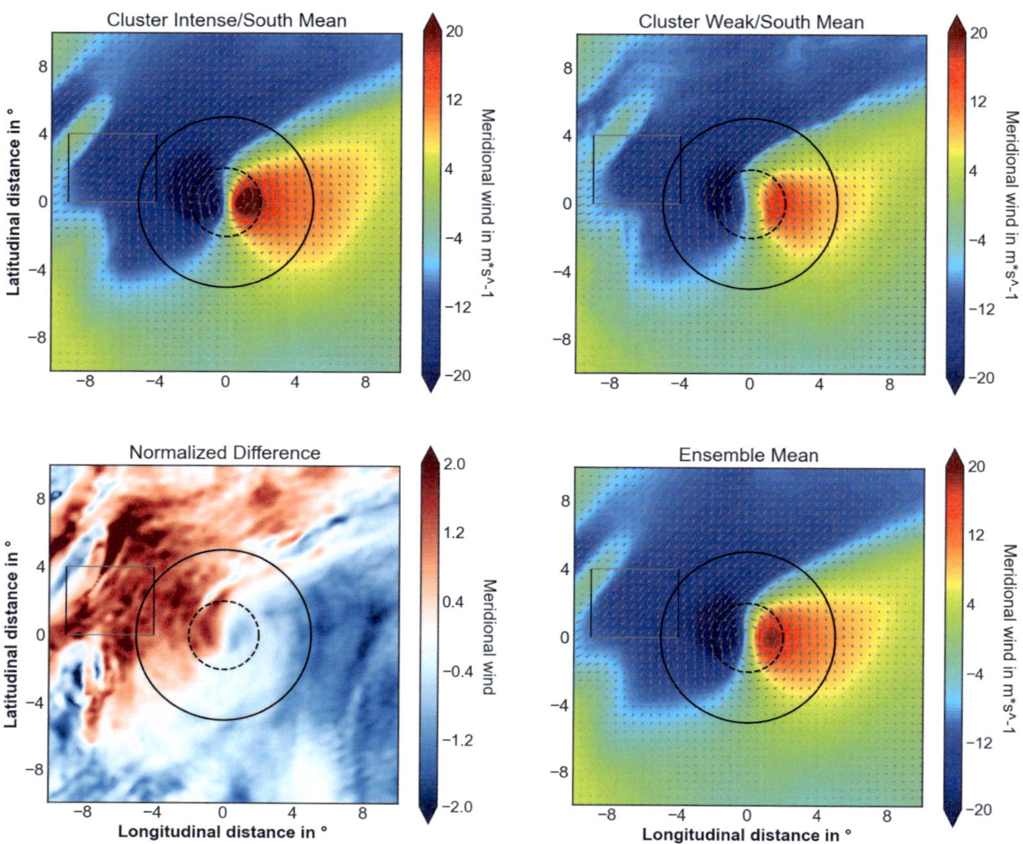

Figure 4.8: Meridional wind and stream field at 925 hPa for TC Atsani at 60 h forecast time. The black dashed line indicates the center and the inner storm area. The area between the black and black dashed line demonstrates the outer bands of the storm. The gray solid line shows the area for mean meridional wind computation for cluster intense (top-left), cluster weak (top-right), normalized difference (bottom-left), and the ensemble mean (bottom-right).

a fixed location is chosen. For the case study, Atsani is Typhoon Goni present in the SCS, while Atsani moves to the northern part of the SCS; the typhoon has a significant influence on the wind field and falsifies the cold surge index. Consequently, a storm-relative approach is used in this study. For every step, the mean meridional wind is calculated in a $5^\circ \times 4^\circ$ cubic 4° to 9° west and 0 to 4° north from the storm center (Fig.4.8). A temperature criterion is not appropriate in this case because the most substantial influence from the cold surge is between 54 and 60 h at a time when the cyclone is located over the northwestern part of the WNP. Subsequently, the wind surge moves for a long time over the relatively warm ocean surface, and this interaction contributes to the problematic identification of a surge event through a temperature consideration. Orographic effects are taken into account for the computation of the index. Therefore, meridional wind over land is not considered. There are two reasons for this. First, large parts of the land are above 925 hPa. Second, the enhanced water vapor fluxes into the cyclone result from a forced thermodynamical disequilibrium between the relative cold and dry air in the lower troposphere and the warm, moist sea surface (Tang and Emanuel, 2012), contributing to TC intensity changes. Cyclones moving over land are not affected by this process. Thus, the disregard of the land surface leads to higher accuracy in the cold surge index.

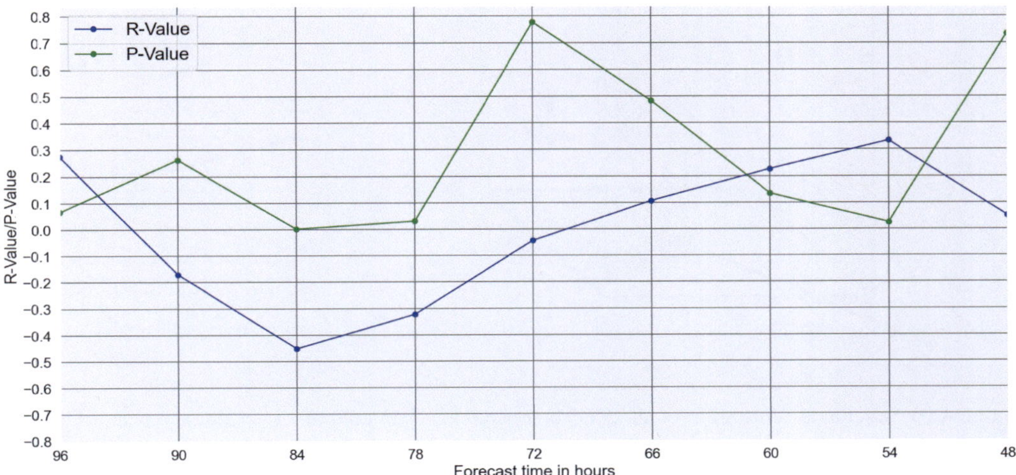

Figure 4.9: Correlation coefficient (r-value) and statistical significance (p-value) between the meridional wind at the 925 hPa pressure level for the time steps between 48 and 96 h and the minimum pressure at 120 h for TC Atsani.

Before the cold surge index is applied to the individual case studies, a correlation analysis is used to examine the relationship between the cold surge and intensity change in the corresponding TC. Therefore, the correlation coefficient is computed between the TC intensity after 120 h and the northerly wind at 925 hPa for different forecast time steps. The index is then applied at the time step with the highest correlation coefficient and a p-value below .05. (Fig.4.9) indicates that for TC Atsani, the highest positive correlation and a p-value below .05 are yielded at 54 h forecast time. Consequently, the strongest influence on intensity after 120 h due to the cold surge is observed here. The time range between 48 and 96 h results from an analysis of the surge onset and duration for the different members (Fig.5.10). Chan (2005) defined characteristic threshold values for the occurrence of such an event: a threshold value of 8 m s for the mean meridional wind at 925 hPa over the considered area (Fig.4.8) indicates a cold surge onset.

4.5 Pressure tendency equation

The PTE equation according to Fink et al. (2012) considers a vertical column from the surface to an upper boundary at pressure p_2 , which is dependent on the geographical location (Fig.4.10). This value is established at 100 hPa for studies of extratropical cyclones in the mid-latitudes and polar areas; in higher latitudes, the stratospheric dynamics are negligible. Knippertz and Fink (2008) show that in southwestern Africa, significant stratospheric dynamics are typical up to 10 hPa. Therefore, the upper boundary is determined in this work after analyzing the pressure variability

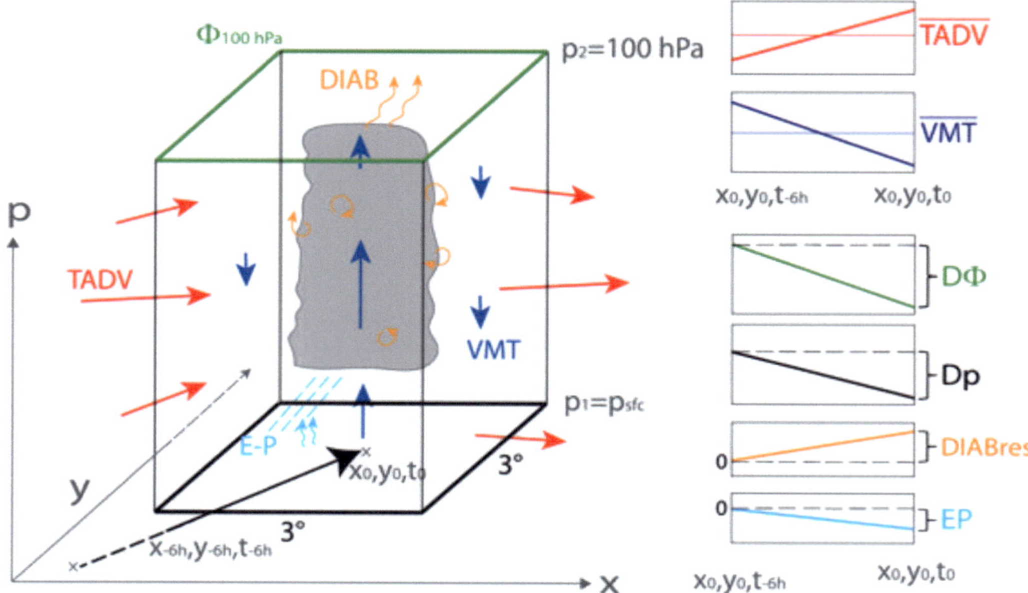

Figure 4.10: Schematic illustration of the PTE computation. The bold arrow in the x-y plane illustrates the motion of the center of a surface cyclone during a period of 6h. The surface pressure tendency equation is evaluated for the $3° \times 3°$ latitude-longitude box extending from the surface to the upper boundary centered on the storm's position at the initial time. The terms of (Eq.4.5), TADV (horizontal advection; red arrows), and VMT (vertical advection; dark blue vectors) are calculated by integrating over the box volume and then averaging over the period as schematically demonstrated in the two graphs in the top-right corner. The computation of the terms $D\phi$, Dp, and DIAB (diabatic processes curled orange vectors), and EP (evaporation minus precipitation; curled blue vectors and dashed blue lines) is indicated in the lower four graphs on the right-hand side. While $D\phi$ and Dp are simple differences between instantaneous values over six hours, EP is the difference between two parameters accumulated between the period. DIAB_RES is the residuum of (Fink et al., 2012, Eq.4.5)

in the stratosphere for the corresponding storm location and seasonal time. The PTE is described by:

$$\frac{\partial p_{\text{surface}}}{\partial t} = \rho s_{\text{sfc}} \frac{\partial \phi_{\text{p2}}}{\partial t} + \rho_{\text{sfc}} R_{\text{d}} \int_{sfc}^{p_2} \frac{\partial T_{\text{v}}}{\partial t} dlnp + g(E-P) + RES_{\text{PTE}} \tag{4.5}$$

or

$$DP = D\phi + ITT + (E-P) + RES_{\text{PTE}}. \tag{4.6}$$

where p_{sfc} is the surface pressure, ρ_{sfc} is the surface air density, ϕ_{p2} is the geopotential at $p2$, R_{d} is the gas constant of dry air, T_{v} is virtual temperature, and g gravitational acceleration. The first term on the left side describes the pressure tendency DP, given as the sum of the geopotential $D\phi$ at the upper boundary p_2, the vertically integrated temperature tendency ITT, and the mass loss due to precipitation or increase by evaporation (E-P). Additionally, a residual term RES_PTE for discretization is taken into account. Surface pressure falls if the upper boundary level decreases under the assumption that the other terms are constant (Fink et al., 2012). Consequently, at the upper boundary, mass evacuation leads to divergent winds. Warming of the column contributes to the expansion and, therefore, mass loss if the column height remains constant. The Integrated Temperature Tendency (ITT) term includes a vertical motion (VMT), a temperature advection (TADV), and a diabatic heating term. The last of these is determined from the residuum of the (TADV) and (VMT) terms (i.e., the sum of physics (PHY) and humidity (HUM)). Latent heat release due to phase changes of water, radiative heating, sensible heat fluxes, and boundary layer turbulence leads to a temperature increase of the column and are stated as (PHY) (Papavasileiou et al., 2020). In contrast, the impact of diabatic heating on humidity is provided by (HUM). Specific humidity decreases due to the condensation of water vapor. Column net warming contributes to a decreasing air density, mass loss, and pressure fall under the condition that the upper boundary of geopotential remains constant. According to Papavasileiou et al. (2020), the decomposition of the (ITT) term indicates large and opposite contributions from horizontal (TADV) and (VMT) over the north Atlantic; these (VMT) and (TADV) terms show a negative correlation because column net warming through warm air advection (TADV > 0) corresponds with upward motions and adiabatic cooling (VMT < 0).

According to Pinto et al. (2005), standard cyclone detection is applied to identify the six-hour positions of the TC. The surface pressure change is evaluated over a grid box of $3^\circ \times 3^\circ$ centered on the TC center. All terms are computed as box area or volume-averaged values. The TADV and VMT are calculated by integrating the box volume and then averaging over the six-hour period. The box in (Fig.4.10) moves along the track of the TC during this period.

5 Case Studies: Over-forecasting

This chapter analyses the possible reasons for the stronger predicted intensification of a cluster compared to a proportion of members showing intensification close to the observational values. Two different case studies in the WNP and SCS are used: TC Atsani over the northwestern part of the WNP and the northern part of the SCS and Typhoon Kammuri over the WNP. Environmental conditions like vertical wind shear, atmospheric moisture, or divergence are investigated. The cold surge index is also applied over the SCS to investigate the impact of cold surges on TC intensity in this area. Ventilation of dry air and the strength of enthalpy and water vapor fluxes are examined.

5.1 Tropical cyclone Atsani

Of the five different clusters of TC Atsani, cluster Intense South (IS) and cluster Weak South (WS) are compared because their cluster centers show a substantial pressure deviation over 25 hPa after 120 h (Fig.4.5). In contrast, the location indicates only minimal discrepancies under 0.1° for the latitude and under 0.7° for the longitude (Tab.4.1). Therefore, the mean pressure of the considered cluster throughout 120 h describes the evolution of intensity and indicates the time step at which the two different clusters show the most robust deviation in their pressure tendencies. Additionally, the standard deviation exhibits the dispersion of the considered cluster.

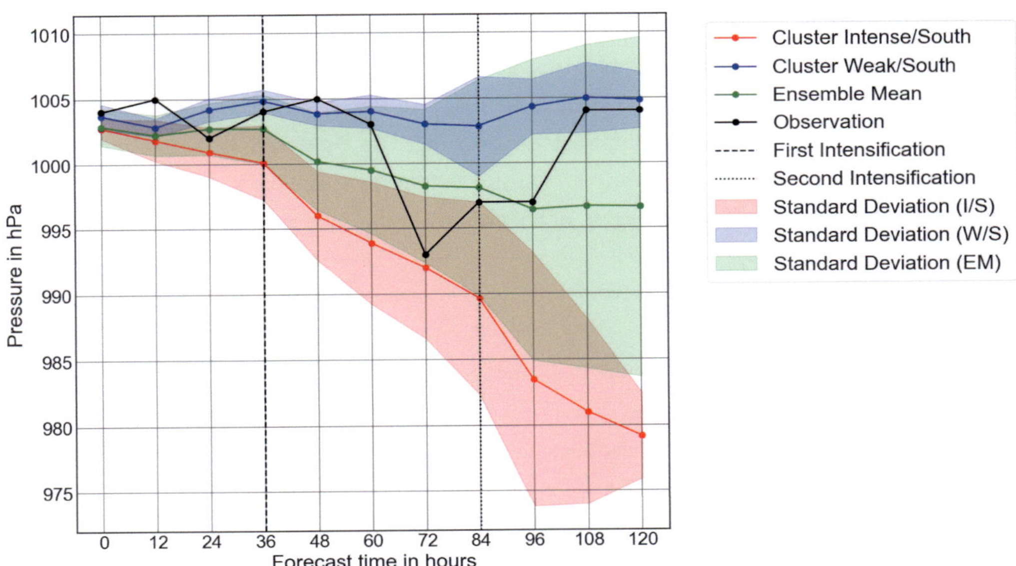

Figure 5.1: Time series of minimum pressure for TC Atsani for cluster Intense/South (IS;red), cluster Weak/South (WS;blue) and ensemble mean (green). The corresponding standard deviations are shaded.

(Fig.5.1) shows that the first spread between IS and WS is identified after 12 h. While the mean values of the clusters develop in parallel up to 12 h forecast time, they change thereafter. Cluster IS demonstrates a small increase in intensity in the following 24 h, while the weak cluster indicates a decrease in storm intensity over the same period. The variations within the cluster remain, especially for the smaller weak cluster up to 36 h. After 36 h, the two clusters diverge substantially for the first time. While IS shows an intensification, WS indicates only slight intensity differences. After 48 h, the standard deviation increases for both clusters and the whole ensemble. (Komaromi and Majumdar, 2015) find that greater ensemble standard deviations are associated with larger mean errors, especially for forecasts with a forecast time under seven days (168 h). Therefore, the uncertainty in the ensemble and the discrepancies from observation rise rapidly while the storm tracks illustrate an anticyclonic turning (Fig.4.4). A second intensification process is shown after 84 h for IS, while WS remains stable in its intensity after this time step. As a result, the largest difference in the 12 h pressure tendency is between 84 and 96 h forecast time. At 96 h, IS demonstrates the highest variability in its intensity. Consequently, the second intensification period indicates a substantial discrepancy in the cluster before the spread decreases to a much smaller extension toward 120 h forecast time. This variability is related to the fact that four members in the cluster passed over land between 96 and 108 h (Fig.4.4). This causes a reduction of the heat fluxes from the surface.

The ensemble mean is closer to cluster WS, and its spread rises rapidly. Therefore, a substantial part of the members is closer to the weak cluster than the strong one. Furthermore, the observation is close to WS up to a forecast time of 60 h. In contrast, IS demonstrates an enhanced deviation from the observation. After 60 h forecast time, the observation shows a rapid increase in intensity and the spread for the weak cluster rises for a short time, while the spread for IS at 72 h forecast time is minimal and rapidly increases after that time. In contrast, IS demonstrates an enhanced deviation from the observation between 36 and 60 h and between 84 and 120 h forecast time. In addition, the spread in the whole ensemble rises rapidly after 36 h. These results imply a first critical point for analysis at 36 h and a second critical point at 84 h forecast time, just before the second, more intense pressure fall of IS and the weak pressure increase for WS. In the following sections, the environmental conditions, especially at and just before these critical points, are analyzed to understand the reasons for the high degree of uncertainty in the ensemble forecast.

5.1.1 Dynamic factors

(Fig.5.2) illustrates only slight differences in the mean environmental deep-layer vertical wind shear between the clusters up to 72 h forecast time. Between 12 and 24 h, the mean value for IS is slightly below that of WS. Both clusters demonstrate a pressure decrease (Fig.5.1). The mean vertical wind shear drops from $15.1\,\mathrm{m\,s^{-1}}$ to $11.1\,\mathrm{m\,s^{-1}}$ for the weak cluster and to $10.3\,\mathrm{m\,s^{-1}}$ for the intense cluster. The extension of the spread for the ensemble showed no clear differences. The cluster IS indicates a higher standard deviation and more members with lower values: 45.4 % of the values are under $10\,\mathrm{m\,s^{-1}}$ and 9.0 % under $8\,\mathrm{m\,s^{-1}}$. (Shu et al., 2013) have shown that a critical value for TCI is $10\,\mathrm{m\,s^{-1}}$ over the WNP. This finding indicates that nearly half of the members in IS provided favorable conditions for TCI. After the first intensification stage, the mean value for cluster intensity

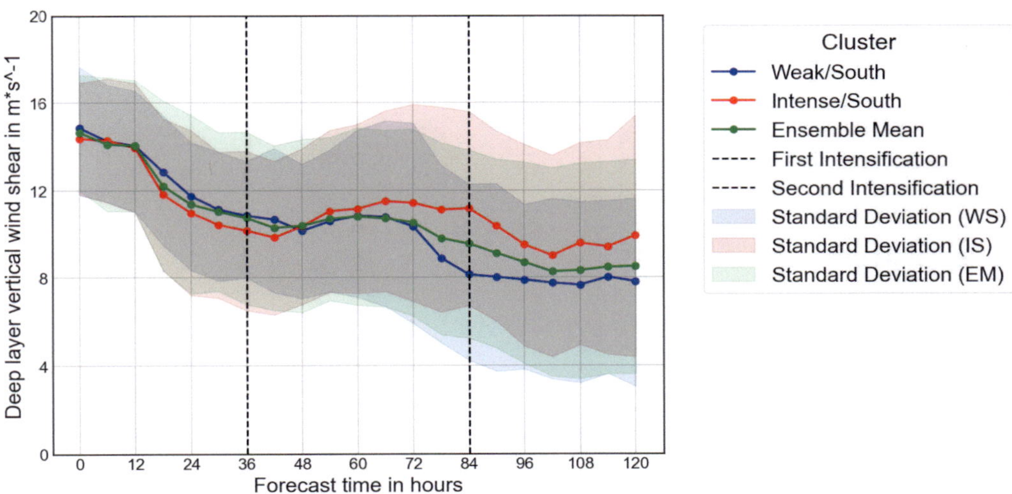

Figure 5.2: Time series of mean environmental vertical wind shear for TC Atsani for cluster Intense/South (red), cluster Weak/South (blue), and ensemble mean (green). The corresponding standard deviations are shaded. The environmental vertical wind shear is the mean value in the outer band of the storm in an annulus between 3 and 5° from the cyclone center.

rises to a value close to $12\,\mathrm{m\,s^{-1}}$. In contrast, the weak cluster value displays only a small increase and drops to a value of $8.2\,\mathrm{m\,s^{-1}}$ after 72h. However, as (Fig.5.1) illustrates, these simulated tropical cyclones indicate no intensification. This effect is confirmed by several investigations that have shown that a low vertical wind shear below $10\,\mathrm{m\,s^{-1}}$ is not sufficient for intensification (Gray, 1998). In contrast to the weak cluster, the intense cluster shows more preferable conditions for intensification at the second critical point: the latter illustrates a rapid intensification while the former shows a slight pressure increase. Consequently, other variables influence TC intensity more substantially. As a result, the environmental vertical wind shear is for TC Atsani for the second intensification stage, no crucial cause for OFC. After the first intensification stage, the intensity spread is to a small extent provided by the deep-layer vertical wind shear. However, the interaction with low- and mid-level core moisture leads to a more substantial effect on the central pressure (Frank and Ritchie, 1999). Consequently, it is essential to consider the moisture at different heights in the simulated cyclones.

5.1.2 Thermodynamic parameters

Several studies have shown that a TC sustains its intensity (i.e., in a high value of the vertical wind shear) much better with a very moist lower and middle troposphere (Zhang and Tao, 2013; Wang and Hankes, 2016). Komaromi and Majumdar (2015) found that the ECMWF model simulates the genesis of TC if the vertical wind shear indicates tremendous values of $15\,\mathrm{m\,s^{-1}}$, but only if the associated waves and their environment are very moist. (Fig.5.3) illustrates the spatial distribution of relative humidity in the lower mid-troposphere at 700hPa for the different clusters. At 48h, a layer of dry air with relative humidity values below 40% to the west of the TC is visible for cluster IS, cluster WS, and the ensemble mean. IS shows lower values to the west and a more considerable extension to the south of the dry layer than WS. The normalized difference indicates a

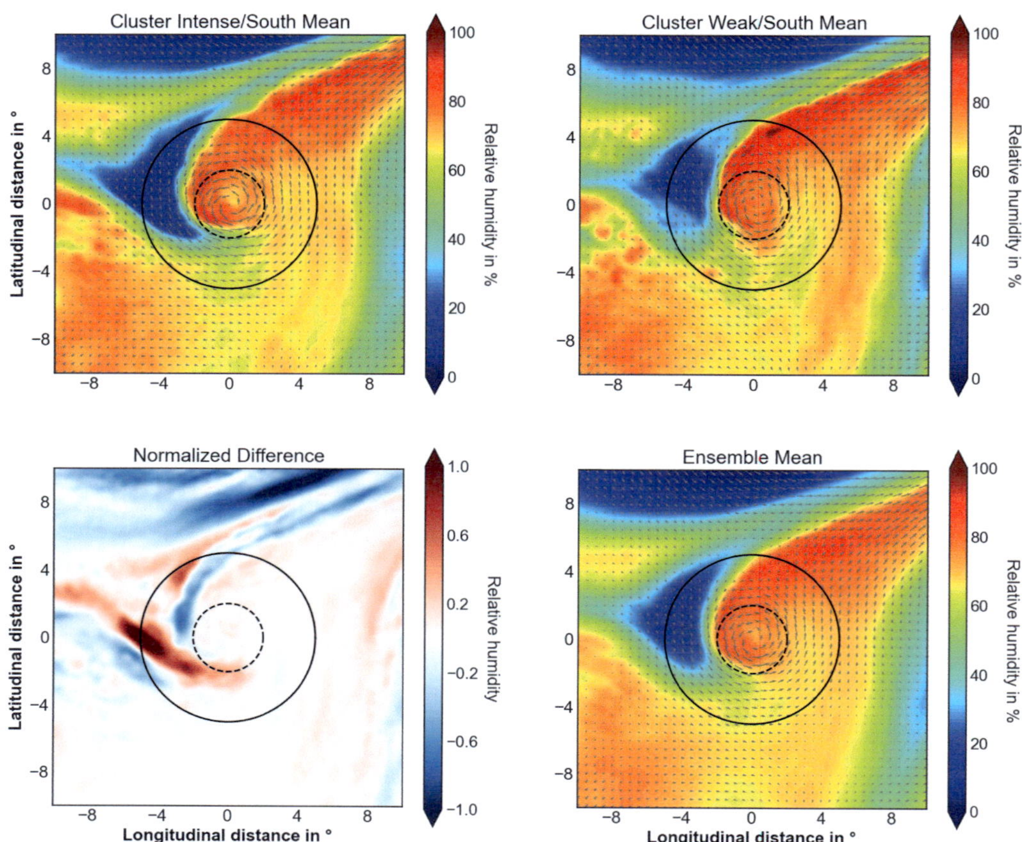

Figure 5.3: Storm-relative composites of mean relative humidity after 48 h forecast time at a pressure level of 700 hPa for cluster Intense/South (top-left), Weak/South (top-right), and all members, showing a storm after 120 h. A calculated normalized difference demonstrates the discrepancies between cluster Weak/South and Intense/South (bottom-left). The wind field is indicated through gray arrows at the corresponding pressure level. The black dashed circle illustrates the center of the cyclone. The area between the black dashed and the solid line represents the outer bands of the TC.

clear distinction between the two clusters to the west and southwest and no apparent discrepancies in the inner core. Additionally, the moist environment of the inner core for IS extends to the west. Consequently, the outward-directed moisture gradient is more intense for IS. Carrasco et al. (2014) investigated the influence of heating rates on the intensity of tropical cyclones. It is shown that less convective activity in the outer cyclone area (associated with lower moisture content) contributes to smaller TCs and a greater intensification rate. At 72 h forecast time, the dry layer is located in the outer band to the south of the cyclone (Fig.5.4); the inner core still indicates no apparent differences between the clusters. Compared to WS, IS illustrates a more westward extended high humidity area and a region with low humidity to the east. This distribution is attributed to the higher deep-layer environmental vertical wind shear for IS. Tang and Emanuel (2010) have shown that the shear contributes to a shift of the deep column moisture to the down-shear side and provides

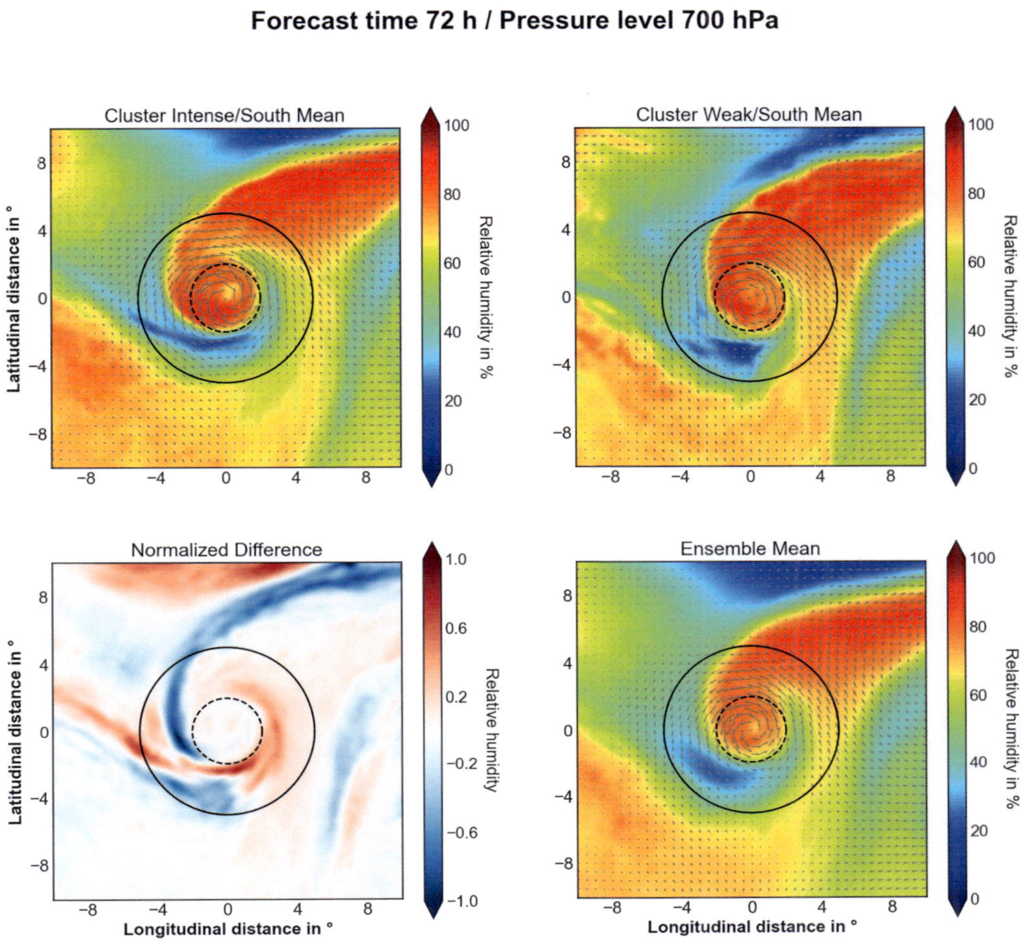

Forecast time 72 h / Pressure level 700 hPa

Figure 5.4: Same as Fig.5.3 but after 72 h forecast time.

a more favorable environment for the entrainment of low-entropy air.

In sum, the inner core in the lower midlevel of the TC shows no apparent differences between the clusters for either of the two intensification stages. Therefore, the only potential signal of a higher intensification in cluster IS is the more intense outward-directed moisture gradient at both critical points, leading to more concentrated convective activity.

The moisture in the boundary layer at 925 hPa indicates a normalized difference close to -1 in the center of the storm after 48 h forecast time (Fig.5.5). At the same time, the area around the inner circle exhibits a positive value close to 1. Subsequently, a much higher value of specific humidity is observed in the inner core and, at the same time, a stronger outward-directed moisture gradient in the boundary layer. (Wang (2012); Wang and Hankes (2016)) have highlighted the importance of a moist boundary layer in the inner core for column moistening and increased convective activity as part of TC intensification. This column moistening includes positive feedback between increased convection, latent heat release, and intensified secondary circulation, strengthening the water vapor

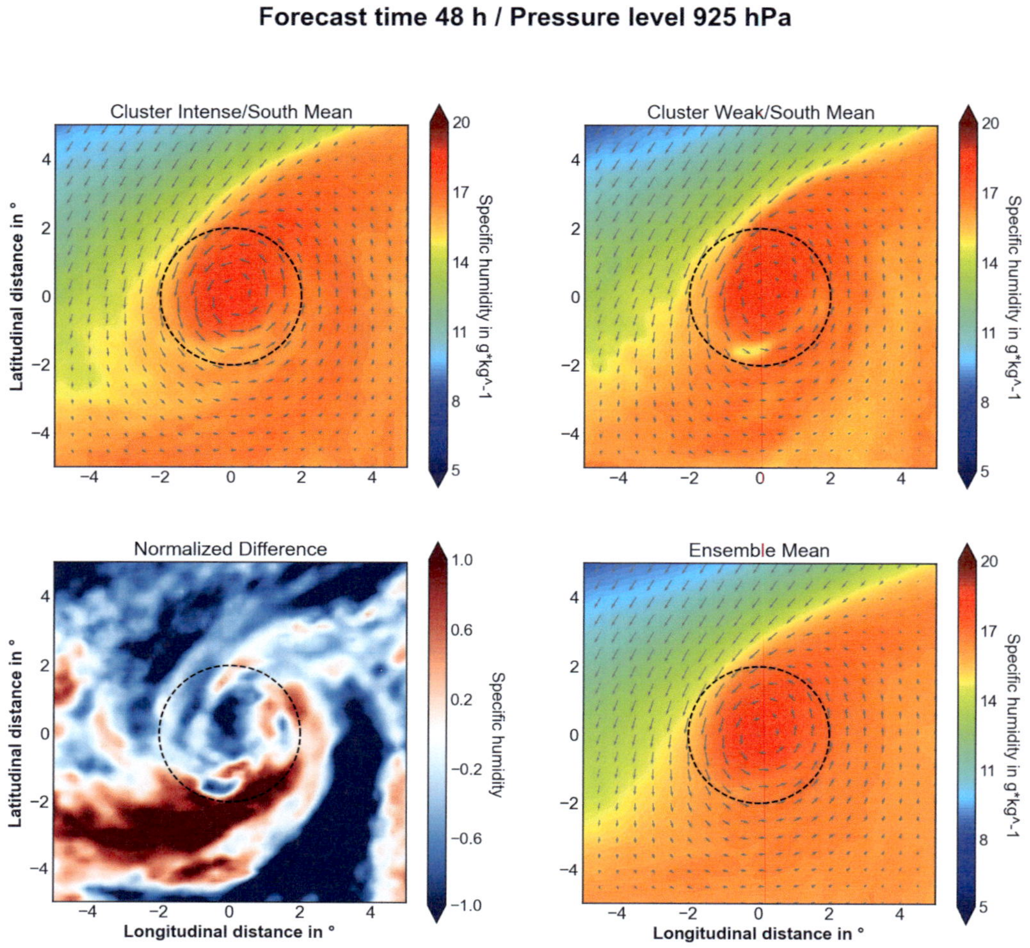

Figure 5.5: Same as Fig.5.3 but for specific humidity at 925 hPa.

fluxes and forcing high moisture (Raymond and Kilroy, 2019). Furthermore, the layer in the outer bands to the north and east indicates normalized differences close to -1. This negative value demonstrates a moister layer for IS compared to WS. The layer to the east feeds the inner core with moisture. At 72 h forecast time, the inner core moisture normalized difference between the clusters was weaker than at 48 h (Fig.5.5). Additionally, a high normalized difference close to 1 is reported to the south and the east. These positive values reflect that the more restricted inner core moisture contributes to more central convective activity in IS. This convective activity in the core area leads to latent heat release, which results in higher efficiency in TCI (Smith and Montgomery, 2016a). However, this evolution of convective activity is strongly influenced by midlevel dry air intrusions. As such, the ventilation index, which describes the strength of low-entropy air intrusions from the midlevel, is analyzed in the following section.

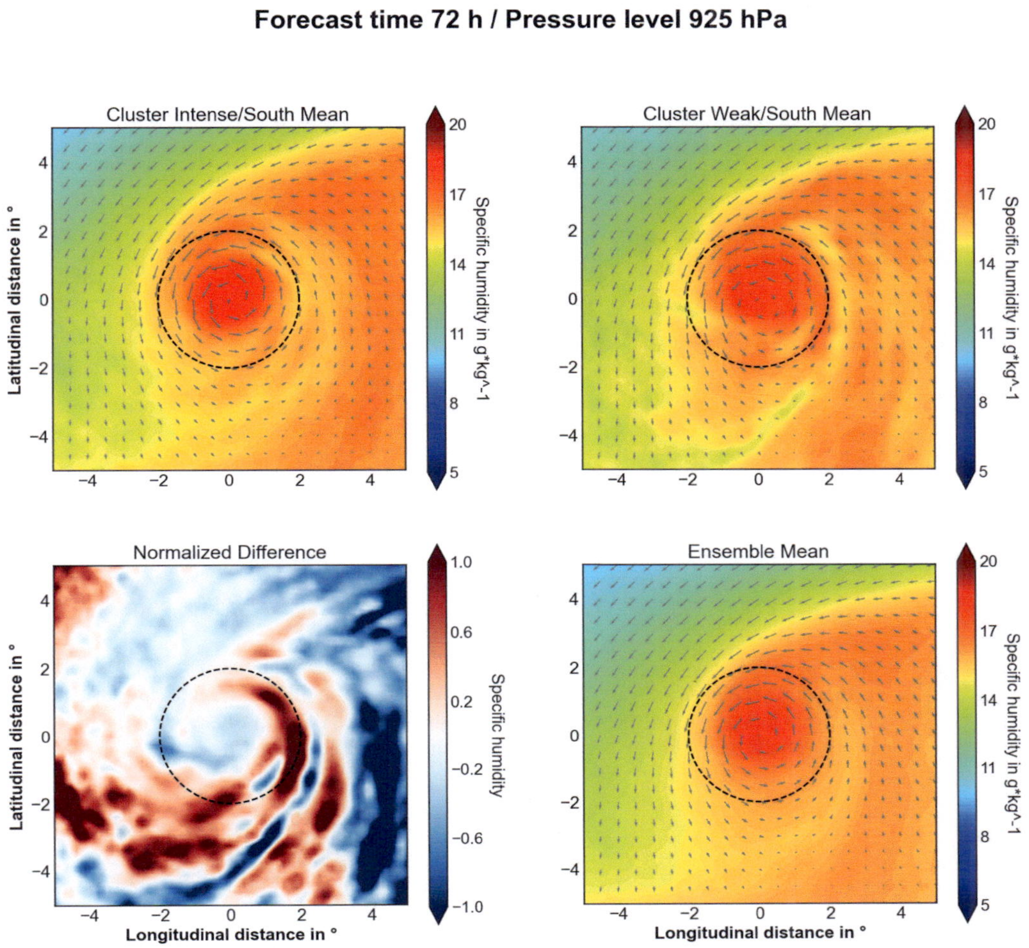

Figure 5.6: Same as Fig.5.3 but for specific humidity at 925 hPa at 72 h forecast time.

5.1.3 Ventilation of dry air

The ventilation index includes three terms: vertical wind shear, entropy deficit, and maximum potential intensity (Sec.2.3.1). The denominator of (Eq.2.6) describes the thermodynamic disequilibrium between the ocean surface and the lower troposphere, while the nominator defines the moisture difference at 600 hPa between the inner core and the environmental area of the cyclone (for further explanation, see Sec.2.3.1).

Both the mean entropy deficit for IS and the ensemble mean between 12 and 36 h are lower than the mean for WS (Fig.5.7). These discrepancies increase after 12 h, amounting to 0.09 at 24 h forecast time. The higher predicted entropy deficit results from a larger midlevel moisture difference combined with enhanced enthalpy fluxes. However, values below 0.7 for the weak cluster are still preferable for TC development (Tang and Emanuel, 2012). Between 42 and 78 h forecast time (i.e., just before the second intensification stage), the deficit for IS is higher than for WS. The mean value closely follows the ensemble mean. After the second intensification stage, the cluster WS and the ensemble mean peak values were above 0.6 and 0.7, respectively. Both display a maximum

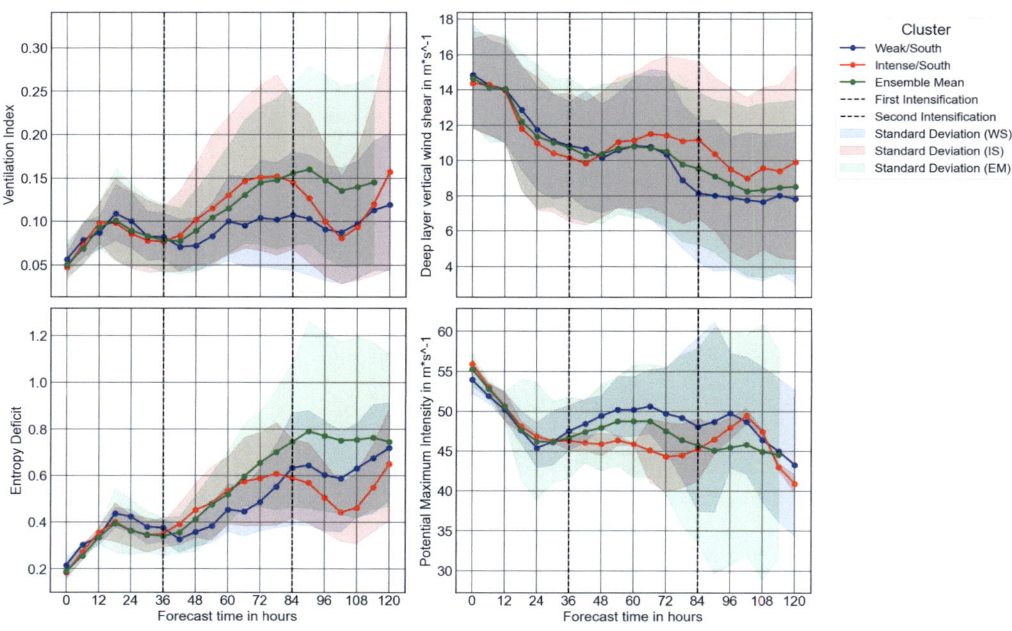

Figure 5.7: Mean values of the ventilation index and their terms with increasing forecast time. The ventilation index (top-left), deep layer vertical wind shear (top-right), entropy deficit (bottom-left) and potential maximum intensity (bottom-right). Shaded areas are the corresponding standard deviations. Black dashed lines indicate the intensification stages.

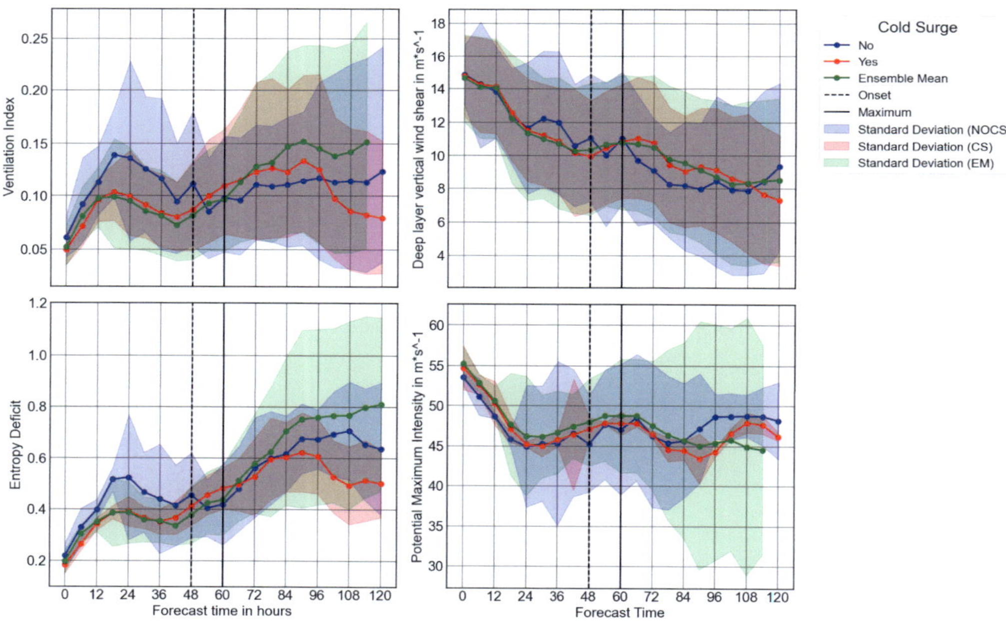

Figure 5.8: Same as (Fig.5.7) but for cluster no cold surge (blue) and cold surge (red). The dashed black line indicates the time of cold surge onset. The solid black line shows the mean cold surge maximum.

value at 90 h close after the second intensification. In particular, the ensemble mean illustrates an enhanced spread. Values over 1 are recognizable. These high values are mainly related to lower

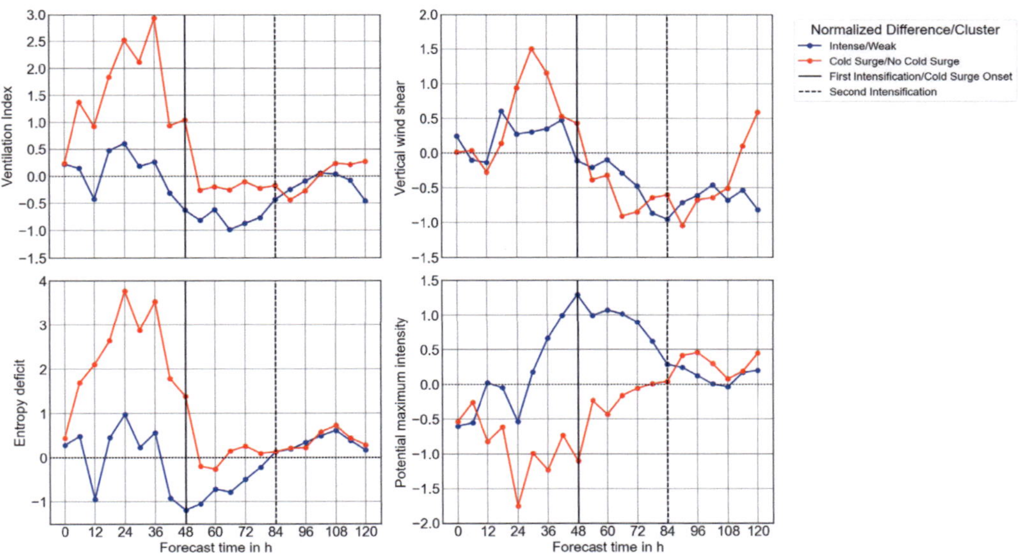

Figure 5.9: Time series of mean normalized differences over a circle with a radius 2° from the storm center for the ventilation index (top-left), entropy deficit (bottom-left), potential maximum intensity (bottom-right), and vertical wind shear (top-right) in an outer circular area between 2° and 5°. The red line indicates the difference between cluster cold surge and no cold surge. The blue line indicates the difference between IS and WS. The solid black line illustrates the first intensification and the cold surge onset, while the dashed black line show the second intensification stage.

sea surface temperatures for the more northward moving storms in the ensemble. The decreased sea surface temperatures contribute to weaker enthalpy fluxes. Consequently, the denominator term in the entropy deficit equation (Eq.2.6) drops.

In contrast, the cluster IS decreases after 84 h to values under 0.5. Therefore, a wide spread between the mean values of the different clusters is illustrated. This spread implies higher midlevel moisture and a lower one in the environmental area for the intense cluster at this time. A strengthening of the secondary circulation follows from that midlevel moisture gradient that fosters tropical cyclone intensity (Wang and Toumi, 2019). (Fig.5.8) demonstrates an enhanced entropy deficit close before the surge onset for cluster No cold surge (NOCS) with weak mean meridional winds to cluster Cold surge (CS) with strong mean meridional winds. These differences are visible up to 54 h forecast time. After this point, there is first a slight and after 96 h a reverse relationship between the clusters. Up to 60 h, all members for both clusters show values under the threshold of 0.7. A small proportion of the NOCS members indicate a value over 0.7, 24 h before the cold surge onset, while all the members of the CS cluster are clearly below this threshold. Thereafter, a much larger moisture gradient appears at the midlevel between the environment and TC center because the boundary layer enthalpy fluxes show no strong discrepancies between the clusters.

The maximum potential intensity is close between the two clusters, as is the ensemble mean up to a forecast time of 36 h (Fig.5.7). An increase is identified for WS after 36 h at the first intensification stage; a second maximum of $50.9\,\mathrm{m\,s^{-1}}$ is reported at 66 h. IS shows a much lower value between the two critical points than WS; a clear distinction between the clusters is only visible between 36 and 84 h, where the intense cluster shows a $4\,\mathrm{m\,s^{-1}}$ to $5\,\mathrm{m\,s^{-1}}$ lower mean value. Due to a more northward location, these inconsistencies are attributed to the limited surface temperatures of a higher number of members in the intense cluster. The larger value of the ensemble mean is also associated with lower surface temperatures. Likewise, a 2 K lower sea surface temperature mean value contributes to a visible difference in the potential maximum intensity of the simulated TC. The NOCS cluster possesses a much lower value between 24 and 48 h (Fig.5.7). The CS cluster shows at the second maximum at the cold surge maximum time a mean value of $47.1\,\mathrm{m\,s^{-1}}$, while the NOCS cluster indicates a second peak value at 96 h forecast time of slightly below $50\,\mathrm{m\,s^{-1}}$.

As a result of the higher entropy deficit together with a stronger spread for WS between 12 and 36 h, the ventilation index shows a higher value than IS at slightly over 0.1; (Tang and Emanuel, 2012) indicate that values over 0.1 are detrimental to TC development and intensification. (Fig.5.7) illustrates this inhibiting factor for 66.6 % members of WS. In contrast, cluster IS presents only 45.4 % of ensemble members with values over the threshold of 0.1. After the first intensification stage, the mean value for WS is close to 0.1, while the value for IS rises steadily up to 78 h forecast time. Subsequently, the ventilation index demonstrates only for the first intensification stage a weak positive contribution to the stronger intensification of IS compared to WS. This positive figure is linked to a combination of weak decreased environmental wind shear and entropy deficit. The maximum potential intensity shows no apparent differences. In contrast, the inconsistencies for the second stage correspond to all three terms. The entropy deficit and the environmental vertical wind shear reflect a normalized difference close to -1 12 to 24 h before the critical point (Fig.5.9). The vertical wind shear is already analyzed in (Sec.5.1.1).

(Tang and Emanuel, 2012) show that TC development is frequently observed for regions with an environmental wind shear lower than $15\,\mathrm{m\,s^{-1}}$, an entropy deficit under 0.7, and a potential maximum intensity higher than $80\,\mathrm{m\,s^{-1}}$. These conditions are fulfilled at 36 h forecast time for the entropy deficit and the vertical wind shear for both clusters (Fig.5.7). At the second critical point, a small fraction of intense members (4.7 %) exceed the threshold for vertical wind shear. Due to the stronger storm intensity in the intense cluster, the simulated cyclones are able, due to their enhanced inertial stability, to survive environmental effects much more effectively (Schubert and Hack, 1982). This resistance is one possible reason for the further intensification of IS after 48 h, despite the detrimental effect of dry air intrusions from the midlevel. For WS, the entropy deficit accounts for 33.3 % of the members with a value above the climatological threshold of 0.7, whereas IS provides only 9.0 %.

In sum, the computed ventilation demonstrates a weak account of pressure fall for the first intensification stage, mainly related to a smaller deep-layer environmental wind shear; the second intensification stage, meanwhile, indicates no contribution to the incorrectly predicted cyclones in the ensemble forecast. This result leads to the assumption that other processes dominate the intensity evolution.

5.1.4 Cold surge

Another reason for the more substantial pressure fall in IS could be a stronger simulated cold surge. To study this effect, the average mean meridional wind over a defined area relative to the corresponding storm center (Sec.4.4) is applied to identify the strength of a simulated cold surge event for the different ensemble members in the clusters. (Fig.5.10) demonstrates the onset, strength, and duration of such an event for four clusters (Tab.4.1). In this work, IS (top-left) and WS (top-right) are compared (Fig.5.10). Evidently, all members show a cold surge event for IS, whereas two members indicate no event for WS. The onset time is similar for both clusters. For WS, all members with a surge event show onset at a forecast time of 48 h; for IS, only four members demonstrate a 12 h onset time. The meridional wind for both clusters is at a forecast time between 60 and 72 h maximal. In terms of cluster number, IS demonstrates more strong events than WS (Fig.5.11) and no member without a surge case. Strong events are defined as cold surge cases with a meridional wind of $-12\,\mathrm{m\,s^{-1}}$ or lower. The maximum wind speed exceeds $-13\,\mathrm{m\,s^{-1}}$ for the strong cluster, but the duration is limited for all members to 24 h. The northern clusters (bottom) demonstrate no apparent differences. However, the southern clusters illustrate that a small portion of WS shows an earlier surge onset.

(Fig.5.11) shows five clusters as well as an additional 'extreme south cluster' and 'extreme north cluster.' All members outside the standard deviation of the latitude are contemplated. Thus, two clusters with seven members for cluster south and eleven members for cluster north are recorded. This calculation illustrates how strongly surges correspond to latitude. No apparent discrepancies between the extreme south cluster and the extreme north cluster are exhibited for the surge events. The southern cluster registers a slightly higher frequency (14.2 %) for strong simulated cold surges than the (northern cluster) but a lower frequency for the moderate cases. The events with no cold surge account for 42.8 % of the southern cluster, while the northern cluster shows only 9.0 %. Consequently, the southern members indicate fewer cold surge events. The main reason for this is that only a small number of members in the whole ensemble simulate a surge case that intrudes into low latitudes smaller than 20° N. One possible explanation for this is that westerly winds block the northerly winds concerning Super Typhoon Goni, which is located over the SCS.

A comparison between IS and WS signals a stronger intensification in IS due to a more intense northerly cold surge (Fig.5.12). Cluster IS includes 36 % of strong and 45 % of moderate cold surge events (Fig.5.11). In contrast, WS displays no strong cold surge events and a higher number of weak surge events. Furthermore, WS provides one member with no cold surge. This result is also demonstrated in (Fig.5.12), which illustrates a stronger intensification after 48 h forecast time for the members with a strong cold surge than those with moderate or weak ones. The corresponding intensification rate after the cold surge maximum is similar for the moderate, weak, and no cold surge clusters: only the strong cold surge cluster indicates a clear intensification after the onset. However, the rate increases after 36 h and then shows a slight decrease after 48 h. The categorization, according to Chang et al. (2005), provides no relevant results, because the different categorized clusters exhibit broad location discrepancies from the SCS (Fig.5.14). Therefore, fixed thresholds contribute to inaccuracies. Consequently, it is essential to apply k-means clustering according to the maximum meridional wind.

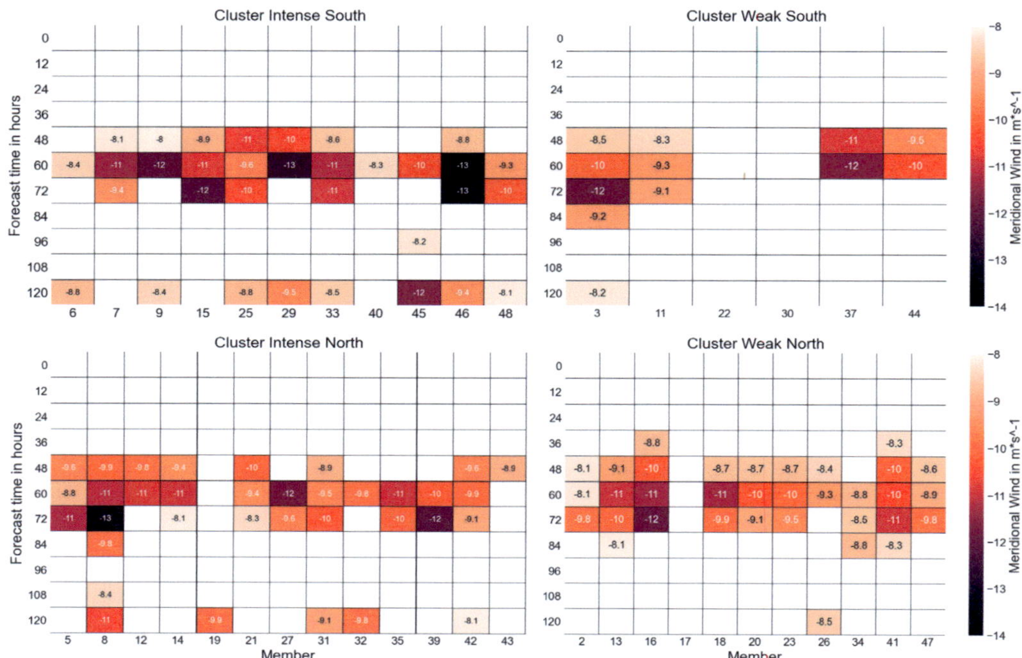

Figure 5.10: Heatmap of mean meridional wind of a $5^{\circ} \times 4^{\circ}$ cubic, 4° to 9° west, and 0 to 4° north from the storm center with increasing forecast time for different clusters. The members with a cold surge event, defined by a mean meridional wind above $8\,\mathrm{m\,s^{-1}}$ at a specific time, are shown in color.

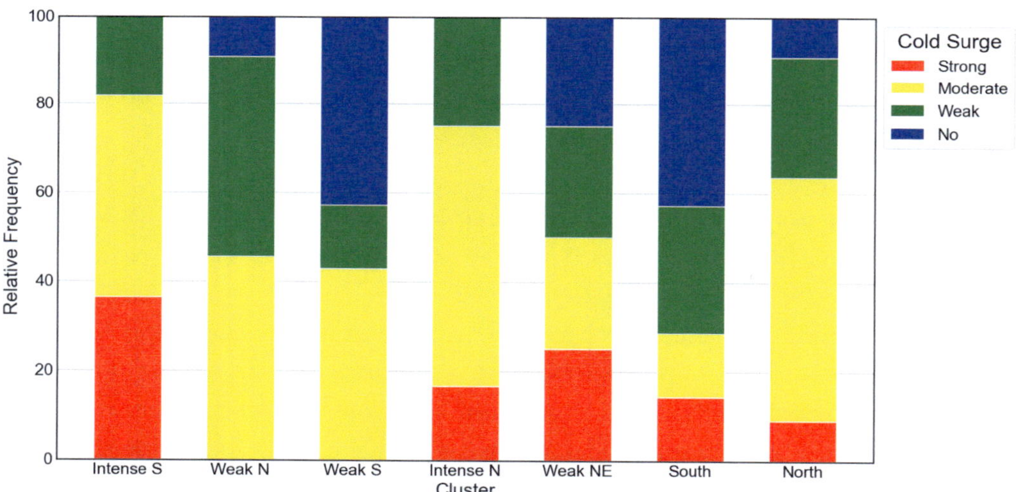

Figure 5.11: Relative frequency of simulated cold surge events for five different clusters (Sec.4.1), an extreme south and an extreme north cluster, categorized according to Chang et al. (2005) 4.4. Red blocks illustrate strong, yellow moderate, green weak and blue events with no cold surge.

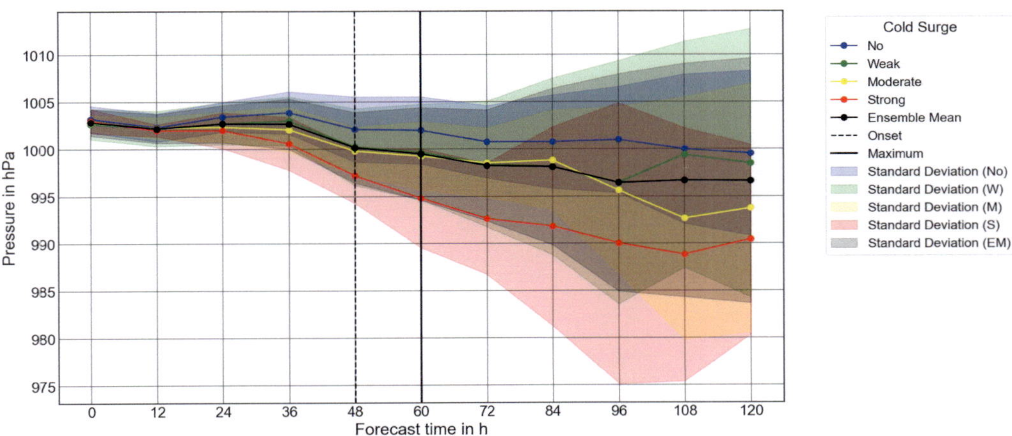

Figure 5.12: Mean pressure tendency and standard deviations of cold surge clusters according to Chang et al. (2005), categorized by maximum intensity. The red line indicate strong, yellow indicates moderate and green indicates weak surge events. The blue line illustrates all members without a surge event.

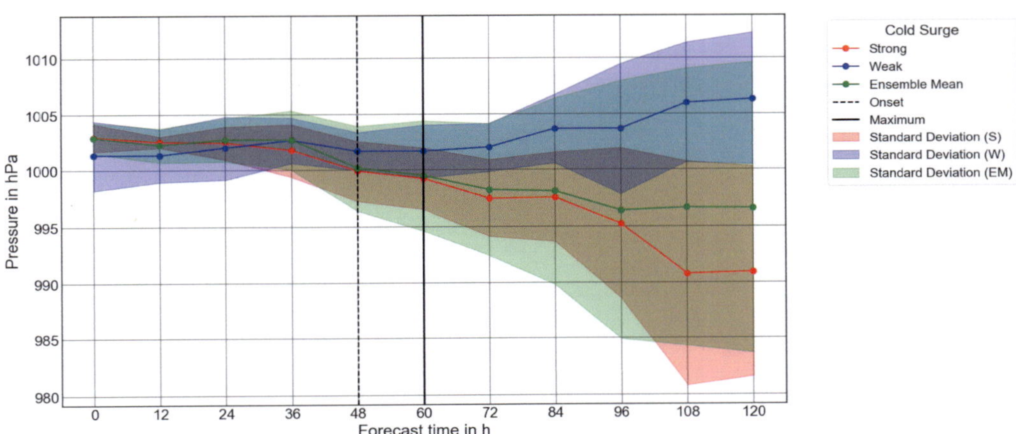

Figure 5.13: Mean pressure tendency and standard deviations of clusters according to the meridional wind. Red line indicate strong cluster. Blue line demonstrates weak cluster. The shaded areas show the corresponding standard deviations.

The members are clustered according to their maximum meridional wind speed using the k-means clustering method (Sec.4.3.1) to differentiate between intense surge events and weak events. Thus, the two clusters with the highest meridional wind difference and the lowest location distance are compared. The intense cluster possesses a mean meridional wind of $-11.46\,\mathrm{m\,s^{-1}}$, while the weak cluster shows a mean value of $-5.02\,\mathrm{m\,s^{-1}}$. The location differences are under 2° (Fig.5.12). The intensity change for the cluster with a strong meridional wind shows a weak intensification after 48 h, while the cluster with a weak mean meridional wind indicates no intensification. However, the highest intensification rate is indicated after 96 h before the transition in the SCS (Fig.5.15), 36 h after the cold surge maximum (Fig.5.13). Nevertheless, a slight increase in the rate of pressure fall is visible after 60 h; this low increased rate implies a positive effect of a strong cold surge on intensification.

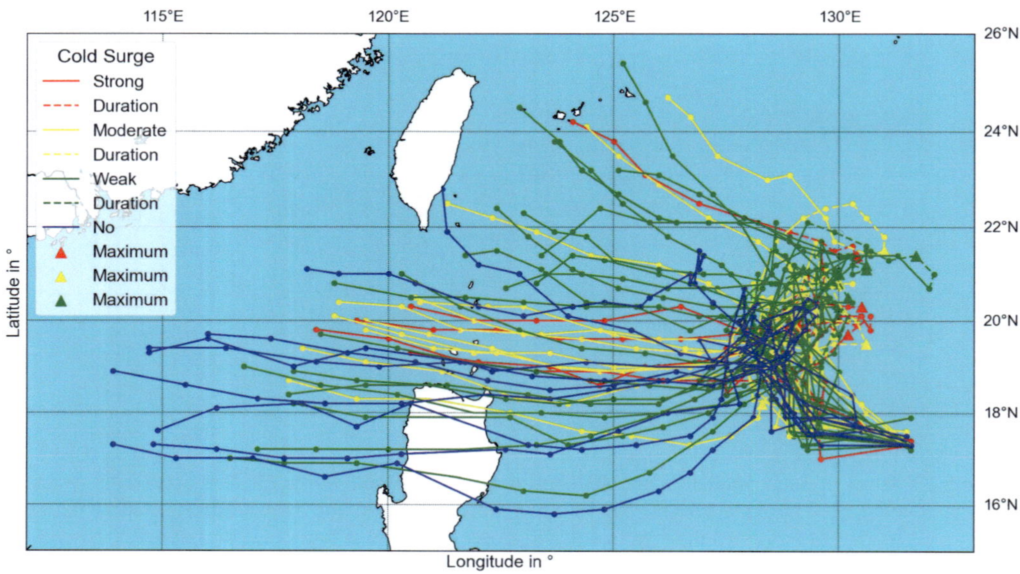

Figure 5.14: Tracks for all clusters categorized to intensity of cold surges. Red tracks indicate strong, yellow moderate, and green weak surge events. Blue tracks all members without and surge event. Dots demonstrate (in 12 h time steps) the location of simulated cyclones. Dashed lines show cold surge duration, with the maximum strength displayed by triangles.

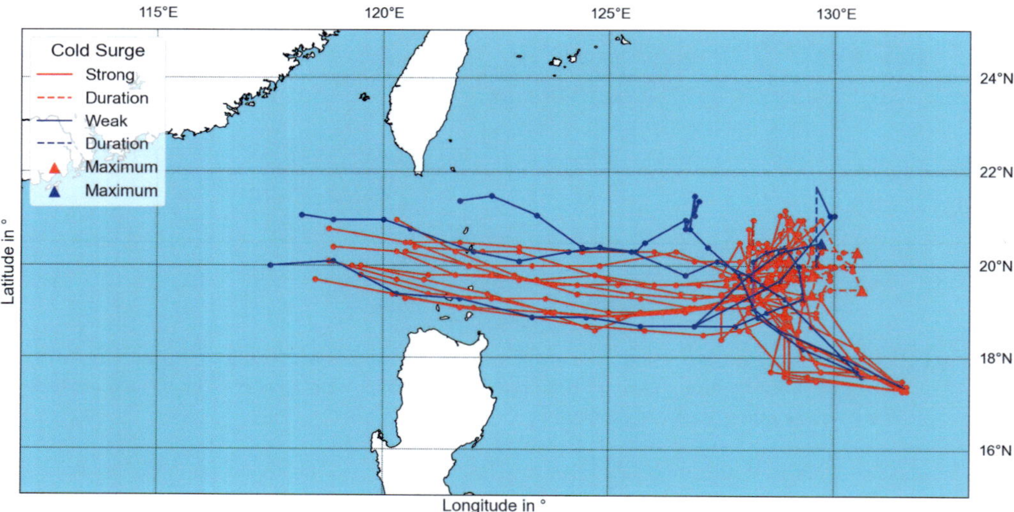

Figure 5.15: Tracks for cluster CS and NOCS categorized to intensity of meridional wind. Red tracks indicate CS. Blue tracks illustrate members of NOCS cluster. Dots demonstrate (in 12 h time steps) the location of simulated cyclones. Dashed lines demonstrate cold surge duration, with the maximum strength displayed by triangles.

5.1.5 Water vapor fluxes

Gray (1998) described that pressure surges increase the pressure difference between the area outside the cyclone and the inner core. This pressure imbalance strengthens the inward moisture fluxes in the lower troposphere and fosters the inner core's moisture convergence. To quantify this process and its influence on TC intensification in some ensemble members, the moisture divergence in the boundary layer and the vertically integrated water vapor flux must be considered.

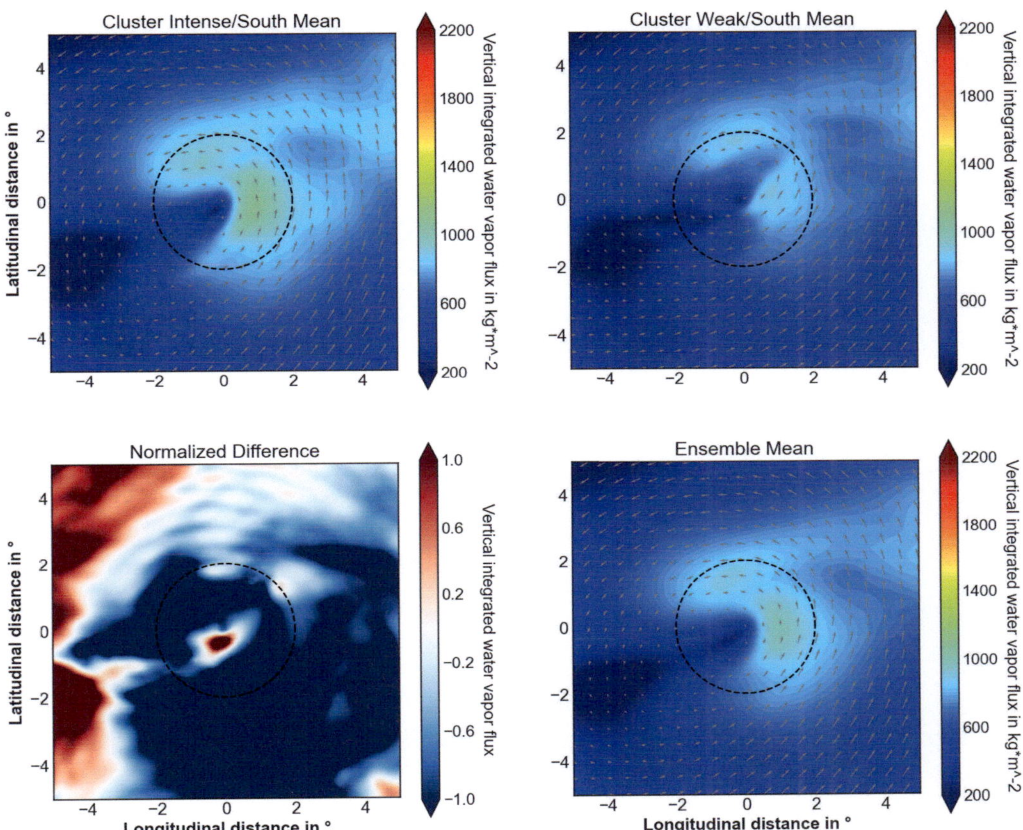

Figure 5.16: Storm-relative composites of mean VIWVF after 12 h forecast time for cluster IS (top-left), WS (top-right), and all members (bottom-right), showing a storm after 120 h. A calculated normalized difference is shown on the bottom-left. The wind field is indicated through black arrows at a pressure level of 925 hPa. The black dashed circle demonstrates the inner core of the cyclone. The area between the black dashed and solid lines represents the outer bands of the TC.

Before the cold surge onset at 12 h forecast time, the Vertically Integrated Water Vapor Flux (VIWVF) already shows a much stronger value over the whole inner part of the cyclone (Fig.5.16). Moreover, the normalized difference demonstrates negative values close to -1 in those outer areas that are less affected by the storm wind field, especially to the south and the east over large parts of the TC environment. For IS, these negative values foreground enhanced VIWVFs as the vital contributor to the more substantial intensification after 12 h in contrast to WS. The highest values above 1,000 kg m^{-2} are identified to the right of the storm translation, where the strongest winds are (top-left). These values are much higher than for WS and also in comparison to the ensemble mean. However, the western area of the outer bands displays positive normalized differences close to 1. This difference describes weaker VIWVFs for the intense cluster. This figure decreases with increasing lead time; after 36 h, a negative value below -1 is recognized (Fig.5.17). This change is possibly related to the cold surge onset since cluster intensity includes more strong cold surge events. According to Murthy and Boos (2018), a large surface enthalpy flux near the circulation center is realized by two factors: an increased wind speed related to the storm intensity and a magnified air-sea enthalpy disequilibrium if the surface winds are constant. In this study, the potential causes that lead to an incorrect intensity prediction are analyzed. Therefore, it is crucial to compare the VIWVF independent of the intensity of the simulated storm. To do so, the normalized difference in the outer cyclone area between 3° and 5° is computed. This region is less affected by the wind field of the TC.

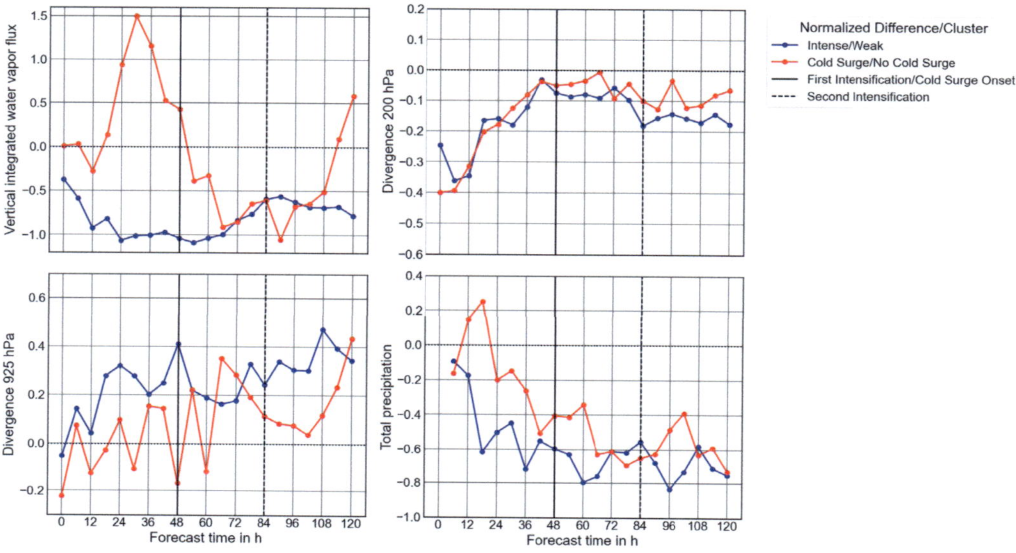

Figure 5.17: Time series of mean normalized differences over a circular area of 3° and 5° for VIWVF flux (top-left), the low-level divergence in 925 hPa (bottom-left), and upper-level divergence in 200 hPa (top-right). The mean normalized difference of total precipitation is averaged over the inner core. A circle with a radius 2° from the center is defined as the inner core. The red line indicates the difference between cluster CS and NOCS, and the blue line indicates the difference between IS and WS. The solid black line illustrates the first intensification and the cold surge onset, while the dashed black line shows the second intensification stage.

The normalized difference of VIWVF in the outer bands between the intense and weak clusters is negative over the whole period of 120 h (Fig.5.17, top-left). It follows that the intense cluster simulates a much more powerful flux than the weak one. Here, the value decreases up to 24 h to -1. At this time, the intensity discrepancies between the two clusters are minimal, and the most intense wind difference is recognized in the inner core and not in the outer bands of cyclones, between the inner core and cold surge area . Additionally, the value rises after 72 h despite the strong pressure reduction of IS and the extended spread between IS and WS. This tendency indicates the influence of an external factor, such as a pressure surge.

(Fig.5.17) demonstrates the effect of the cold surge on VIWVF, low-level divergence, and convective activity as described by the amount of total precipitation (red line). The normalized

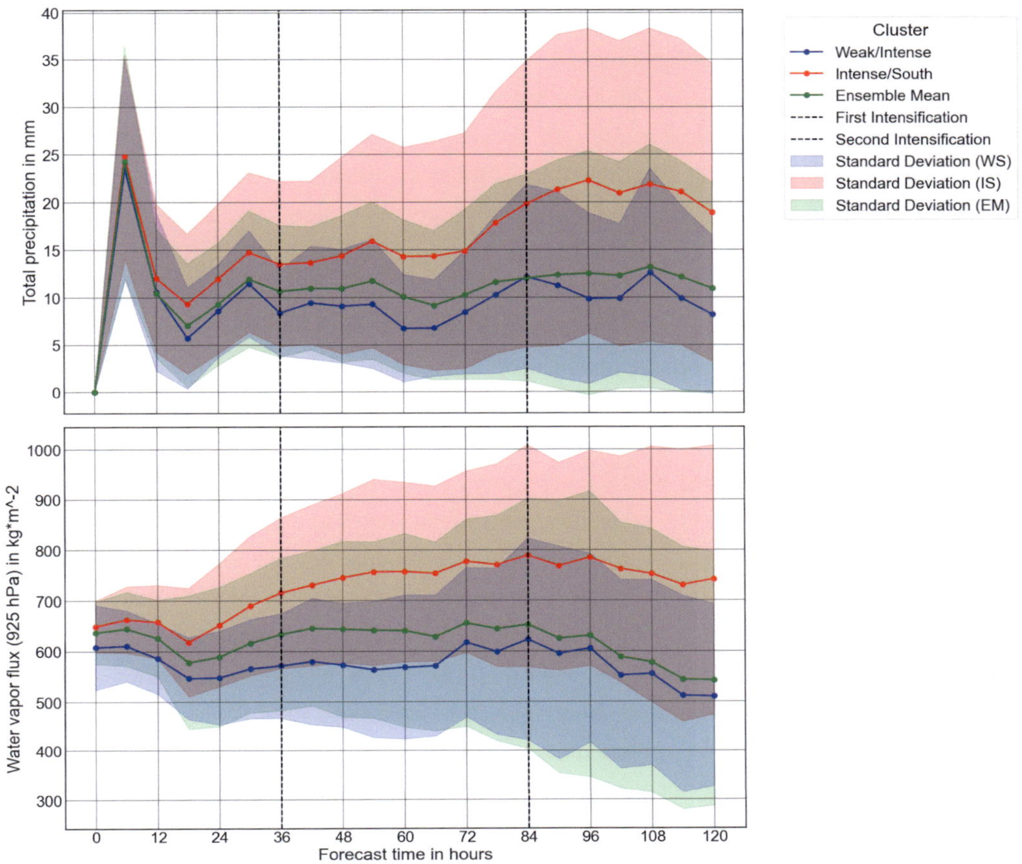

Figure 5.18: Total precipitation and water vapor flux (925 hPa). The red line shows the mean value of cluster intensity, and the blue line shows cluster weak. Shaded areas are the corresponding standard deviations.

difference of VIWVF starts near zero and increases rapidly just before the surge onset. After the onset, the curve decreases to zero; after the surge maximum, it diminishes strongly to close to -1. This tendency highlights the effect of the surge event on the VIWVF. Another proxy responsible for the increased intensification due to pressure surges is the strength of low-level convergence (bottom-left). This variable shows values close to zero up to 60 h. After the surge onset, its value grows sharply to 0.35 and decreases again after 66 h. For the IS and WS clusters, the normalized difference of low-level convergence rises after 12 h to positive values between 0.2 and 0.4 and remains stable except for minor fluctuations. Positive values are related to a stronger low-level convergence for IS. The upper-level divergence shows higher values for IS and for the cold surge cluster between 12 and 24 h (top-right). These differences diminish with forecast time and indicate, especially for the cold surge/no cold surge cluster, only small negative values between -0.2 and 0.0.

The total precipitation is negative for both computed normalized differences (bottom-right). The spread of the mean values between the two clusters indicates an increase after 36 h forecast time. This proxy is strongly dependent on the intensity of the simulated TC due to the positive feedback between enhanced convective activity, leading to a more intense latent heat release and increased total precipitation. This feedback implies that the rising total precipitation of IS in relation to WS and CS in relation to NOCS is associated with the water vapor flux divergence in the boundary layer shift at about 24 h (Fig.5.18). The time series shows a substantial drop in total precipitation for IS and WS. At a forecast time of 36 h, the normalized difference registers the first minimum with a value of -0.7 (Fig.5.17). After this time, it fluctuates between -0.6 and -0.8. For CS and NOCS, a similar tendency is visible, with one exception: the value rises sharply after the onset and declines after the meridional wind maximum at 6 h from -0.2 to -0.6.

In sum, the normalized differences show increased VIWVF for the intense cluster compared to the weak cluster, especially before the first intensification stage. In combination with an enhanced low-level convergence in the outer area, no apparent variations in the upper-level divergence, and increased precipitation, the VIWVF is a contributing factor to an over-forecasted TC. An enhanced low-level convergence combined with a much more powerful VIWVF indicates that a TC has a more intense forced inward water vapor flux, which contributes to a more substantial low-level convergence; this in turn induces vertical motions, convection, and therefore a higher amount of 6 h total precipitation (Emanuel, 2003). One reason for the greater inward water vapor flux is the cold surge, which leads to a pressure imbalance between the outside rising pressure and the inner core pressure minimum (Gray, 1998). (Fritz and Wang, 2014) have shown through numerical simulations that the inward water vapor flux from large radii contributes mainly to a negative enthalpy gradient and increased total precipitation in the inner core. The more powerful water vapor fluxes positively correlate between the primary circulation, secondary circulation, and convection for TC development. Therefore, the enhanced flux of water vapor has a considerable impact on TCI.

5.2 Tropical cyclone Kammuri

Typhoon Kammuri is the second considered storm in the WNP that shows an over-forecasting of TCI. As with TC Atsani, the k-mean clustering method is applied to identify several clusters (Sec.4.3.1). The best choice of cluster number is stated using the elbow method and the calculation of the silhouette coefficient. As a result, four separate clusters with distinct properties were identified. In contrast to TC Atsani, the k-mean clustering is applied at a forecast time of 84 h since this is the point with the second highest mean intensity error and a more extensive spread than at 120 h. Additionally, just before that time, the mean intensity error decreases rapidly, indicating enhanced forecast uncertainty before this critical point at 84 h. Consequently, clustering dependent on the pressure and location at this point is the best choice for further analysis. The two clusters with the largest intensification and the lowest location discrepancies are applied to examine sensitive acting environmental variables. Cluster Intense 1 (I1) and Cluster Weak 1 (W1) show a pressure difference of 24.8 hPa and a negligible location distance. Both the latitudinal and longitudinal distances are under 0.2 hPa.

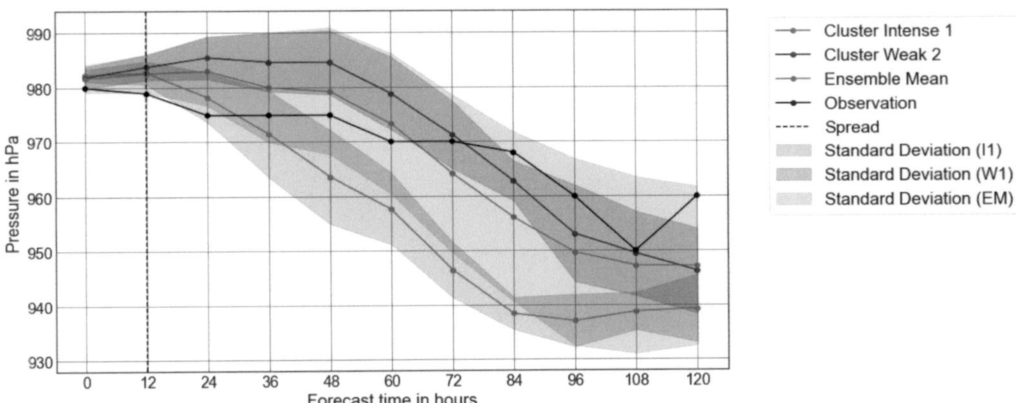

Figure 5.19: Time series of minimum pressure for Typhoon Kammuri for cluster I1 (red), cluster W2 (blue) and ensemble mean (green). The corresponding standard deviations are shaded.

The pressure tendency evolution demonstrates that clusters I1 and W2 show the first intensity deviation after 12 h (Fig.5.19): I1 indicates a continuously sharp increase in intensity, while W2 shows no differences in the mean up to 48 h. As a result, W2 possesses a mean pressure of 995 hPa after 48 h, whereas cluster I1 possesses only a minimum pressure of 964 hPa. This strong increase in the ensemble exposes an enhanced uncertainty. Therefore, it is important to examine the environmental conditions that affect TC intensity in the period between 6 and 24 h, with special attention paid to 12 h forecast time. The observation begins at the lowest point of the ensemble. It shows almost no intensification and is therefore, according to the intensification rate, especially close to W2 between 24 and 48 h. In the period between 48 and 84 h, both clusters show nearly the same intensification rate, which is much higher than the pressure fall in the observation. The strongest deviation of I1 from the observation occurs after 84 h. Additionally, this is also the point with the second highest mean intensity error. The spread maximum peaks after 96 h with a maximum of 13,5 hPa. However, the most intense spread rate is visible between 24 and 36 h. This enhanced error in the ensemble is possibly related to the high deep-layer vertical wind shear, given that previous studies have noted that a large shear magnitude contributes to less predictable TCs, especially regarding the onset time of rapid intensification (Zhang and Tao, 2013; Tao and Zhang, 2014).

5.2.1 Dynamic factors

To examine these assumptions, the deep-layer environmental vertical wind shear is analyzed (Fig.5.20). It displays at the beginning for both clusters, and the ensemble mean has a value of $16\,\mathrm{m\,s^{-1}}$. This high value indicates an unfavorable condition for further TCI. After 12 h, the mean value of I1 decreases, while the value for W2 remains stable. Furthermore, the spread of the ensemble, including the two clusters, increases. This spread expands at a forecast time of 12 h, where the TC's translation speed is very low, and it turns anticlockwise (Fig.5.21). In addition, the storm is influenced by an upper tropospheric trough to the north and is steered within a strongly amplified and fluctuating wind field. This environment is critical to the skill of TC intensity prediction (Komaromi and Majumdar, 2014). During the intensification, mean values of about $13\,\mathrm{m\,s^{-1}}$ for the intense cluster show more preferable conditions, as at the beginning of the period.

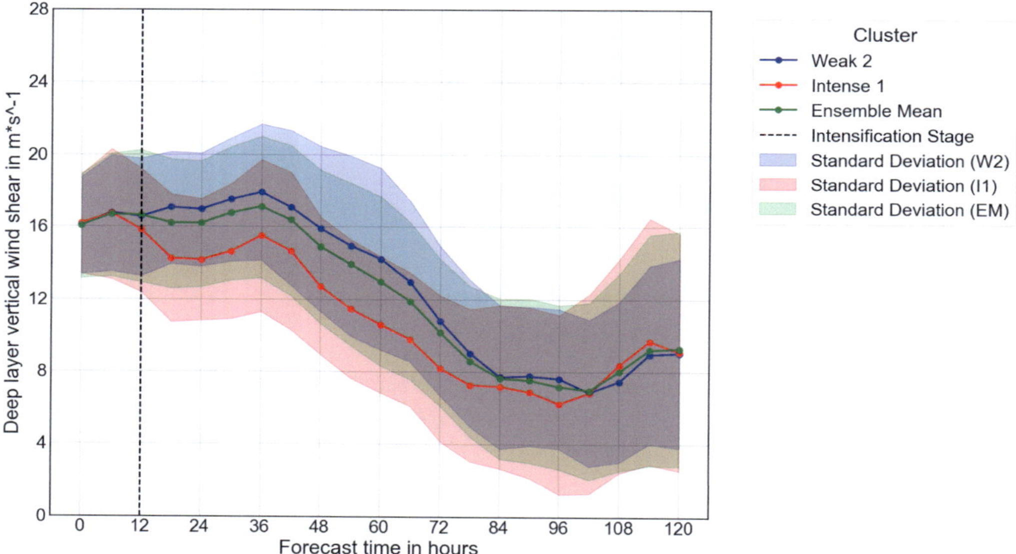

Figure 5.20: Time series of mean environmental vertical wind shear for Typhoon Kammuri for cluster I1 (red), cluster W2 (blue), and ensemble mean (green). The corresponding standard deviations are shaded. The environmental vertical wind shear is the mean value in the outer band of the storm in an annulus between 3° and 5° from the cyclone center.

According to (Gray, 1968)), a deep-layer vertical wind shear above $15\,\mathrm{m\,s^{-1}}$ is an inhibiting factor of further TCI. In contrast, values above $12\,\mathrm{m\,s^{-1}}$ only contribute to intensification if the low- and mid-levels are very moist (Komaromi and Majumdar, 2015), as the enhanced deep-layer shear leads to a vertical misalignment of the midlevel to the low-level vortex. As a result, dry intrusions from the midlevel into the low-level storm center are more probable. With further intensification, the vertical wind shear drops for both clusters, while the weak one shows a later decrease according

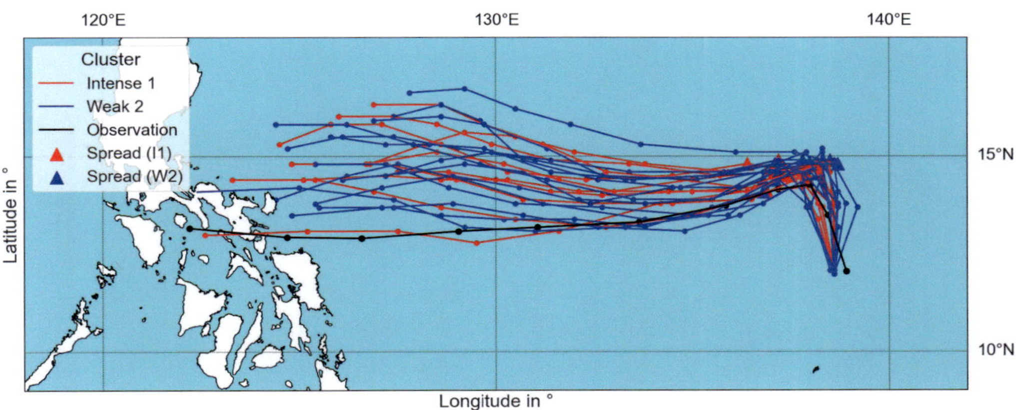

Figure 5.21: Tracks for I1 (red) and W2 (blue). Dots demonstrate (in 12h time steps) the location of simulated cyclones and triangles the location at the critical point.

to their change in intensity. After the wind shear falls below $15\,\mathrm{m\,s^{-1}}$, the simulated TC intensifies. The largest spread after 12h symbolizes the high uncertainty of the location of a trough to the northwest (not shown). Consequently, the deep-layer wind shear contributes to the more significant intensification of cluster I1 than W2. These results are consistent with investigations by Zhang and Tao (2013), who studied the effect of vertical wind shear on TC intensity predictability and demonstrated the lower predictability of TCI in environments with a larger shear magnitude, which forces the convection farther away from the center and reduces the secondary circulation strength.

5.2.2 Thermodynamic Parameters

The storm-relative composites of relative humidity at 12h forecast time show in the inner core no apparent differences between the clusters and the ensemble mean (Fig.5.22). Both clusters and the ensemble indicate a layer of dry air to the west, extending with increasing forecast time to the south. This layer of dry air is close to the cyclone, but more intense for cluster I1 than for W2. The relative humidity shows favorable conditions for both clusters to the north and northeast of the cyclones, with values above 60%. This pattern of high humidity is extended into the tropical cyclone area. Values above 70% are only visible to the west and north in the inner core area. Furthermore, the southeastern outer band demonstrates much lower moisture for I1 than for W2. Humidity values for both clusters locally are below 60%. These values are in a circle area with a radius of 300km from the storm center, a limiting factor for TC development (Komaromi and Majumdar, 2015). Nevertheless, the intense cluster displays a rapid intensification over the following hours.

The normalized difference is close to 1 in the southeastern area (bottom-left). This demonstrates a much lower humidity in the outer bands, which leads (together with a moist core) to a smaller convective activity area that is restricted to the core area (Carrasco et al., 2014). As a consequence, the storm's intensification rate is much higher. Over the following two days, the high moisture area in the lower mid-troposphere develops into a restricted region close to the core. However, the moisture difference between the clusters still indicates only minor discrepancies. To the west of the outer bands, the boundary layer moisture reports a much lower value for cluster I1 in contrast

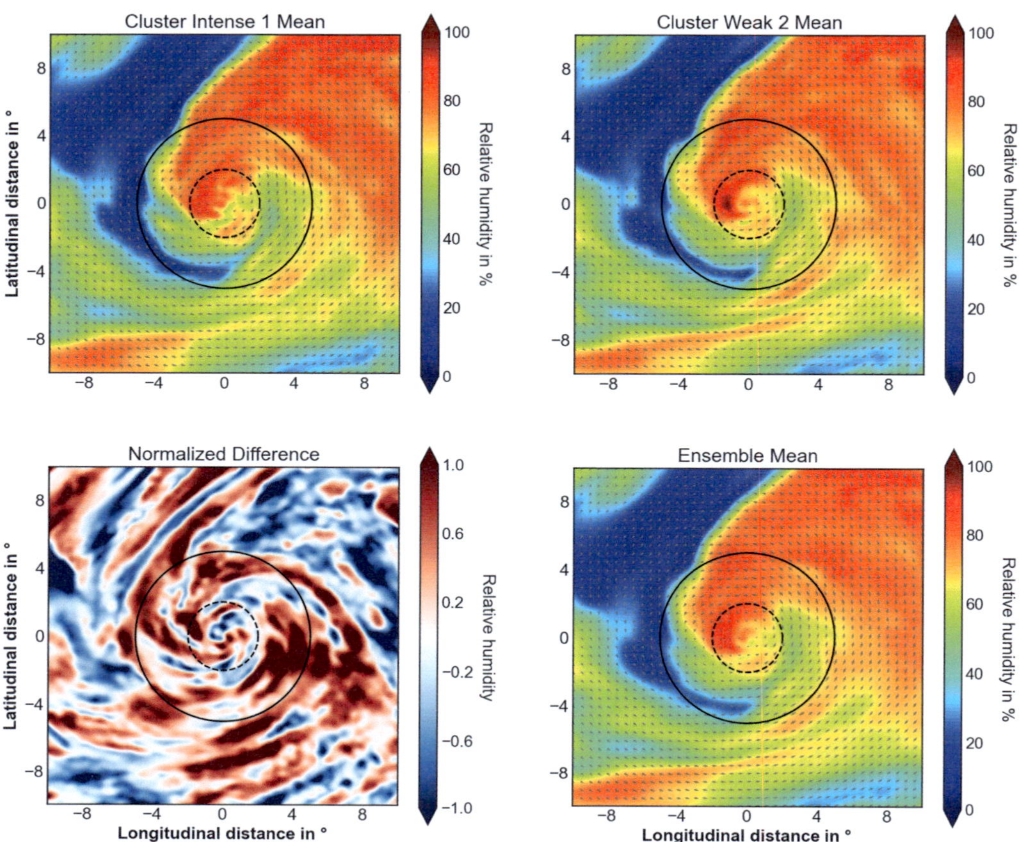

Figure 5.22: Storm-relative composites of mean relative humidity after 12 h forecast time at a pressure level of 700 hPa for I1 (top-left), W2 (top-right), and ensemble mean (bottom-right). The calculated normalized difference is shown bottom-left. The wind field is indicated by black arrows at the corresponding pressure level. The black dashed circle demonstrates the inner core of the cyclone. The area between the black dashed and solid line represents the outer bands of the TC.

to W2 and the ensemble mean. This discrepancy is consistent with the moisture difference in the lower mid-troposphere. However, for the boundary layer-specific humidity, the high normalized differences are extended to the south. The outer bands show values over 1 to the southwest of the storm, while the lower mid-troposphere moisture demonstrates neutral values close to 0. Indeed, it should be considered that relative humidity is strongly dependent on temperature; lower temperatures thus contribute to increased relative humidity if the absolute moisture is constant. However, storm-relative composites of temperatures illustrate only weak temperature differences in the outer bands at 12 h.

In comparison to the midlevel, the moisture in the boundary layer at 925 hPa indicates a reversal sign to the east (Fig.5.23). Much higher values of specific humidity for I1 illustrate a normalized difference close to -1 (bottom-left). This negative value implies enhanced moisture in this area,

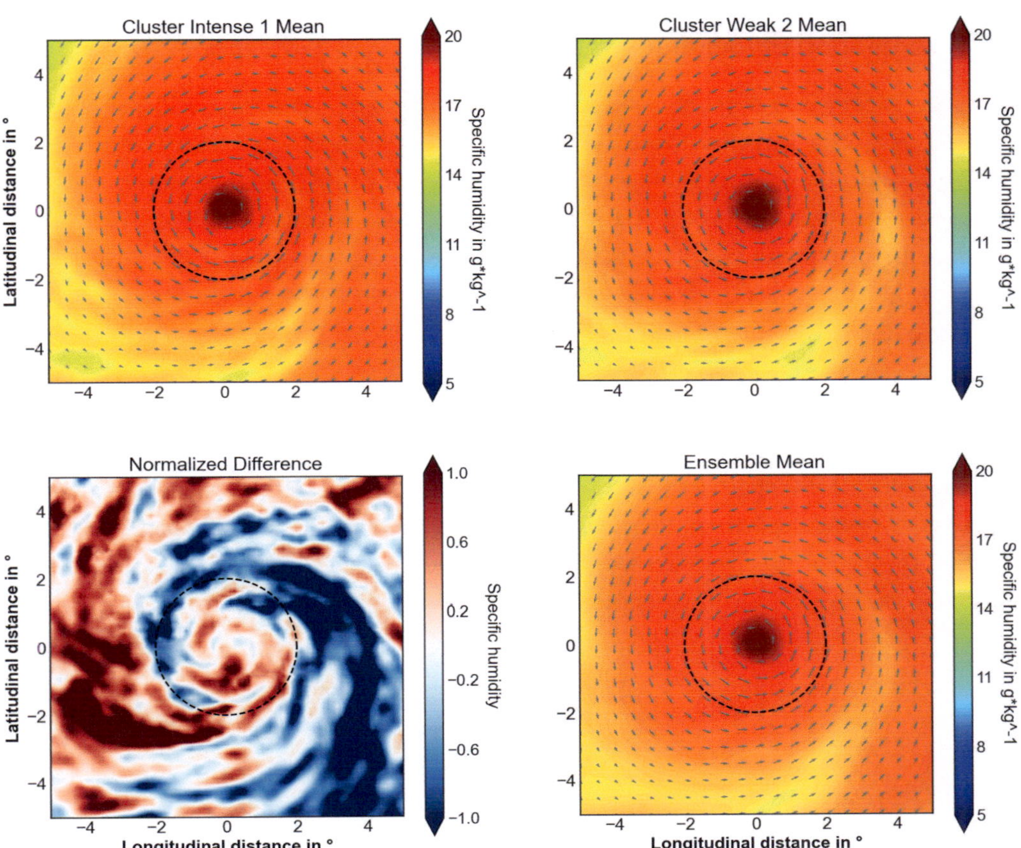

Figure 5.23: Storm-relative composites of mean specific humidity after 24 h forecast time at a pressure level of 925 hPa for I1 (top-left), W2 (top-right), and ensemble mean (bottom-right). The calculated normalized difference is shown bottom-left. The wind field is indicated by black arrows at the corresponding pressure level. The black dashed circle demonstrates the inner core of the cyclone. The area between the black dashed and solid line represents the outer bands of the TC.

intruding from the north into the outer, core area of the simulated TC. Nonetheless, at 12 h forecast time, no apparent discrepancy between the clusters is demonstrated in the center. The outer center area shows especially lower values for I1 to the south. These moisture distributions reflect that the different simulating storms show no inconsistencies in intensity at this time. Nevertheless, the moistening from the east into the inner cyclone area (together with a drier air layer to the west) signals increased intensification for I1, in contrast to W2. The resulting outward-directed moisture gradient in the boundary layer forces the secondary circulation and contributes to pressure fall. Ge et al. (2013) note that midlevel dry air in the right quadrant down-shear area of a developing TC is detrimental to intensification. For the simulated storms, this region is located to the north or northwest of the storms, where relative humidity at 700 hPa is above 70 % for both clusters. This humid air layer provides a more sustainable storm system against the high values of deep-layer vertical wind shear (Komaromi and Majumdar, 2015, Sec.5.20). In turn, the ventilation index and its terms are analyzed in the next section to investigate these processes and their contribution to forecast errors in the ensemble.

5.2.3 Ventilation of low-entropy air

This index is computed from the vertical wind shear discussed in (Sec.5.2.1). It demonstrates a sharp decrease of values below 15 $\mathrm{m\,s^{-1}}$ for I1 just before the increasing spread. Consequently, the deep-layer vertical wind shear shows an influential contribution to the pressure tendency difference between the two clusters up to 48 h. The second term is the entropy deficit (Fig.5.24). This variable describes the relation between the midlevel entropy difference and the strength of enthalpy fluxes. Clusters I1 and W2 demonstrate no apparent differences up to 24 h. The mean values peak at 12 h for the first time at 0.7 and fall over the following six hours below 0.6. The range in the ensemble remains relatively small up to 24 h. It increases strongly after 24 h forecast time for I1. Values close to 1 describe unfavorable values for over 50 % of cluster members. At 36 h forecast time, W2 presents a 0.1 lower value; the normalized difference at that time is a minimum of -0.62. After 30 h, the potential maximum intensity averages close to 40 $\mathrm{m\,s^{-1}}$; the highest value is identified as 46.3 $\mathrm{m\,s^{-1}}$ at the beginning. Both clusters demonstrate a relatively small spread up to 36 h. Additionally, the normalized difference is close to zero, especially at the critical point after 12 h and during the intensity rate inconsistencies up to 48 h.

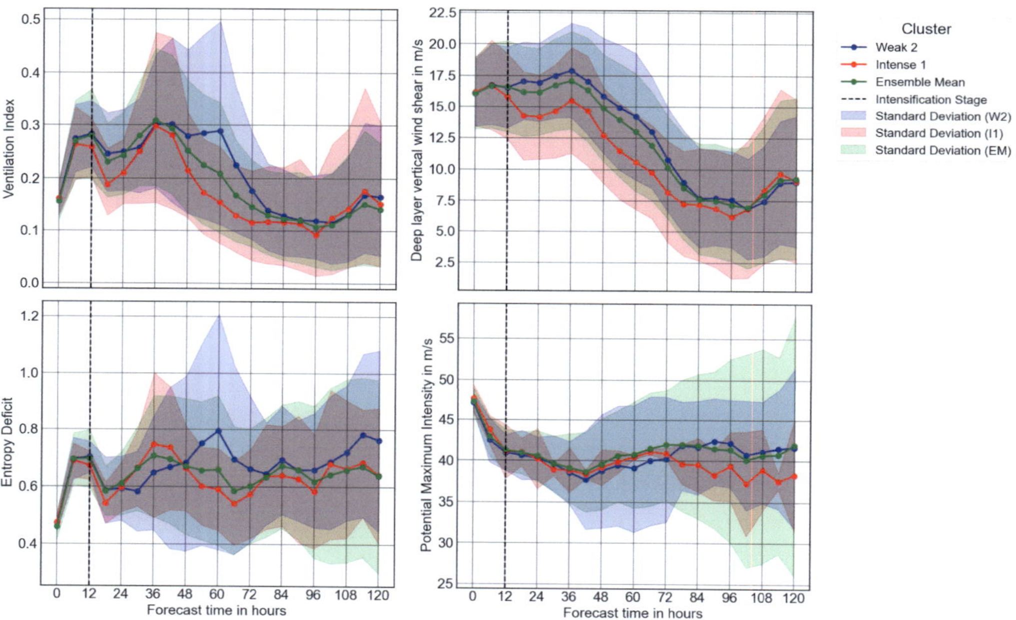

Figure 5.24: Mean values with forecast time for the ventilation index and its terms: VI (top-left), deep layer vertical wind shear (top-right), entropy deficit (bottom-left) and potential maximum intensity (bottom-right). Dashed areas are the corresponding standard deviations.

Consequently, the sharp peak of 1.2 in the normalized differences of the ventilation index is mainly attributable to deep-layer vertical wind shear. This dynamic variable contributes to a much higher probability of dry entrainment from the midlevel into the storm center for W2. However, both

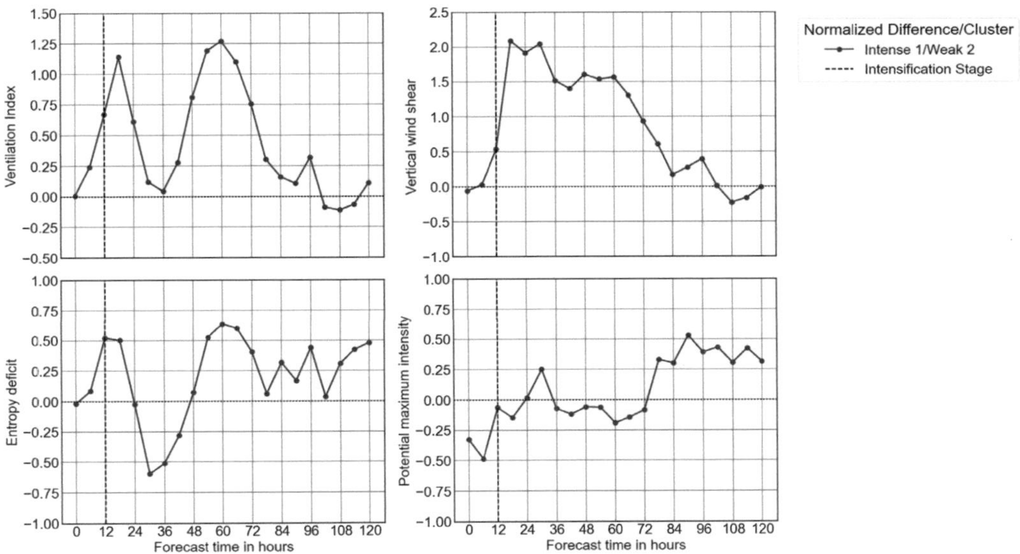

Figure 5.25: Time series of mean normalized differences over a circle with radius 2° from the storm center for ventilation index (top-left), entropy deficit (bottom-left), potential maximum intensity (bottom-right), and vertical wind shear (top-right) in an outer circular area between 2° and 5° between clusters W2 and I1. The dashed black line illustrates the first spread increase.

clusters demonstrate unfavorable values above the threshold of 0.1 for intensification (Tang and Emanuel, 2012). This result is associated with a high deep-layer vertical wind shear and a relatively low potential maximum intensity. The only term that shows suitable values for a high proportion of members at the critical forecast times between 12 and 24 h is the entropy deficit. The second maximum of the normalized difference for the ventilation index is recognized at 60 h forecast time (Fig.5.19). At this time, both clusters indicate an intensification. This peak is mainly due to broad discrepancies in the entropy deficit: W2 shows an average entropy deficit of 0.8, while the mean value for I1 is 0.58 at the same time. In addition, the spread in the weak cluster is significantly enhanced. This divergence implies high uncertainty in moisture prediction. The lower border of the standard deviation is close to the border of I1. However, some members are also included with highly unfavorable values above 1. Nevertheless, the intensity shows an increase for this cluster; evidently, those other parameters overcompensate for this negative effect on intensification.

5.2.4 Water vapor fluxes

One parameter that works against the dry air intrusions in the midlevel is the water vapor fluxes, which supply the low level with moisture. Murthy and Boos (2018) demonstrate that a large surface enthalpy flux near the circulation center is a result of two mechanisms: the increased winds strongly related to TC intensity and an enhanced air-sea disequilibrium if the wind speed is capped at a constant value (Sec.2.3.3). These surface enthalpy fluxes include the sensible and latent heat fluxes. The latent heat fluxes describe the water vapor fluxes.

From these results, the VIWVF is strongly dependent on TC intensity. Consequently, it is crucial to consider this variable just before the spread in the ensemble at 12 h forecast time and outside the core area (inner circle) of Typhoon Kammuri at forecast times with increasing spread between

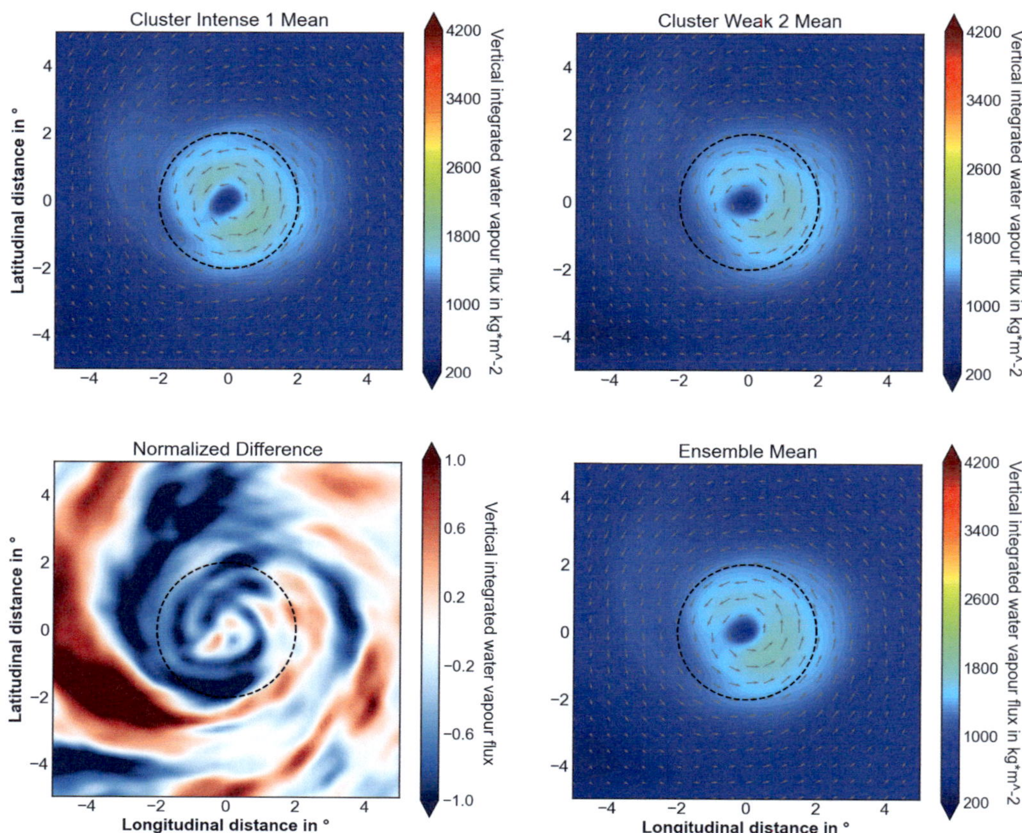

Figure 5.26: Storm-relative composites of mean vertical integrated water vapor flux after 24 h forecast time for I1 (top-left), W2 (top-right), and all members (bottom-right), showing a storm after 120 h. A calculated normalized difference is shown at the bottom-left. The wind field is indicated by black arrows at a pressure level of 925 hPa. The black circle demonstrates the inner core of the cyclone.

the cluster's intensity (Fig.5.26). The core area possesses a radius of 2 and is where an ordinary cyclone shows the strongest winds (McBride, 1995). Storm-relative composites indicate for I1 a higher amount of VIWVF at the western and northern outer bands of the cyclone at a forecast time of 12 h. Normalized differences are close to -1. Outside an area of 3° from the center, a band of high normalized differences demonstrates a stronger VIWVF for W2. In contrast to the band of low values, the area of high values does not reach the storm's inner region. The inner core region shows an enhanced VIWVF for I1, especially to the west and south. To the east, no clear difference between the clusters is visible. Both clusters illustrate maximum values near $2,000\,\mathrm{kg\,m^{-2}}$ to the southeast. In contrast to the ensemble mean, I1 demonstrates higher values to the west and the south. The higher values to the northwest correspond to more intense meridional winds due to an enhanced northeast Passat wind surge (not shown); this wind system acts as a pressure surge event (Gray, 1998). As a result, a strong inflow generated by the enhanced pressure differences between the inner core and the surge-related outer areas induces a

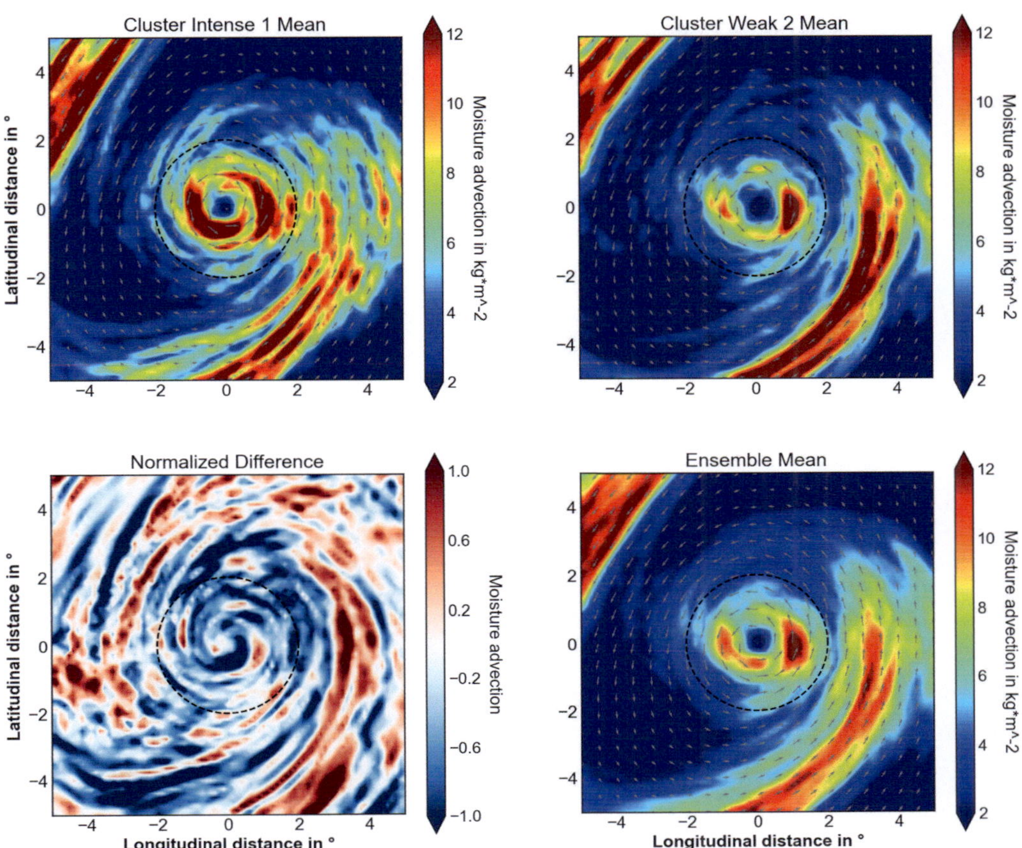

Figure 5.27: Storm-relative composites of mean moisture advection after 24 h forecast time for I1 (top-left), W2 (top-right), and all members (bottom-right). A calculated normalized difference is shown at the bottom left. The wind field is indicated by black arrows at a pressure level of 925 hPa. The black circle demonstrates the inner core of the cyclone.

strengthening of the secondary circulation, which transports water vapor into the inner cyclone area.

Furthermore, the intense cluster indicates a more intense negative radial gradient of VIWVF from the center towards the outer bands of the inner cyclone area. This result implies a fostered convective instability near the circulation center and leads to TC development (Murthy and Boos, 2018). Gao et al. (2019) investigated the role of air-sea latent heat fluxes on TC development. This study states that an enhanced latent heat flux contributes to higher specific humidity in the boundary layer. This process induces positive feedback between a strengthening secondary circulation, more intense inward-directed moisture transport, increasing storm intensity, and tangential winds. However, this case study shows no apparent difference in the boundary layer between the clusters at this time and 24 h thereafter. In contrast, according to Fritz and Wang (2014), the main contribution of enthalpy fluxes to TC development is attributed to the inward moisture flux from large radii. The water budget equation includes two terms that describe the pressure change processes due to water

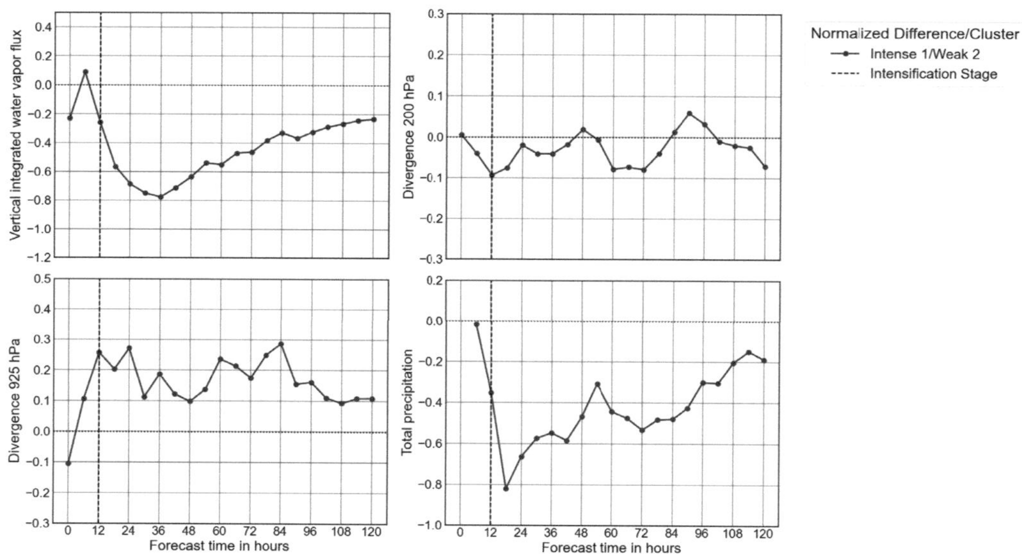

Figure 5.28: Time series of mean normalized differences over a circular area of 3° and 5° for the VIWVF (top-left), low-level divergence at 925 hPa (bottom-left), and upper-level divergence at 200 hPa (top-right). The mean normalized difference of total precipitation is averaged over the inner core. A circle with radius 2° from the center is defined as the inner core between W2 and I1. The dashed black line illustrates the first significant spread increase.

vapor fluxes (Sec.2.3.3). (Fig.5.27) illustrates, at a forecast time of 24 h, an enhanced moisture advection from the east to the north into the inner region of Kammuri. This advection is higher for I1 in the outer area of the storm in comparison to W2 and the ensemble mean. Furthermore, a fostered advection is demonstrated for I1 in the inner core close to the center. W2 shows a more intense transport of moisture around the storm. However, this moisture is not intruded into the storm; it affects only the outer storm bands and the surroundings. These results indicate that a higher amount of moisture transport is present in the inner bands of the storm system for I1 and at the same time a lower amount in the outer bands, especially to the east of the storm system.

For wind divergence, the computed mean normalized difference in the outer bands is contemplated. Here, the divergence indicates a solid rise to a positive value of 0.25 between 0 and 12 h, followed by a more intense convergence for I1 in contrast to W2. Additionally, the VIWF shows slightly increasing values during this period. Consequently, the cluster simulates fostered water vapor fluxes, associated with a more intense low-level convergence, leading to increased rising motions in the inner bands. This process contributes to a higher convective activity, which is also forced by a magnified upper-level divergence. As a result, the model produces a higher amount of total precipitation (Raymond and Kilroy, 2019). The total precipitation demonstrates the highest normalized difference after 18 h, after the peaked upper-level divergence but before the highest discrepancies in VIWVF, low-level divergence, and moisture advection. After 18 h, the negative normalized difference of total precipitation increases to values close to -0.6.

In sum, both storms show an enhanced VIWVF for the intense cluster just before the intensification stage. These discrepancies are broader in the inner core for TC Atsani. However, Atsani indicated at 12 h forecast time a slight intensity difference between the two clusters. Therefore, the cyclone's inner winds are more powerful, and the VIWFV inconsistency is also related to storm intensity. Nonetheless, Atsani shows a more sustainable negative mean normalized difference in the outer bands. This longer duration of stronger VIWFV for the intense cluster corresponds to the cold surge onset at 48 h forecast time. However, the initial enhancement is not associated with the cold surge. This increased VIWFV for the intense cluster provides a strong negative radial gradient in VIWFV from the center to the outer inner area (Murthy and Boos, 2018). This gradient fosters higher convective instability near the circulation center and increases the precipitation rate. In contrast, Kammuri demonstrates a more intense moisture advection from outer areas, and the negative radial gradient does not lead to enhanced low-level moistening for the intense cluster. The spatial distribution of VIWVF normalized differences indicates the connection with strong North Passat wind surges. This wind surge develops at the critical point of 12 h forecast time.

6 Case Study: Under-forecasting

This chapter analyses the strongest under-forecasting event in the WNP between 2011 and 2020 (September to November). The sensitive acting environmental conditions are examined at the critical points. An investigation of the PTE terms for the three storms is presented in the last section. Finally, the influence of the terms on pressure fall is compared for the different events at the critical points.

6.1 Tropical cyclone Champi

The ensemble of the pressure tendency demonstrates a small spread up to a forecast time of 36 h (Fig.6.1). At the initial point, the observation indicates a 5 hPa higher pressure than the ensemble mean. After 120 h, the observed pressure is 52 hPa lower than the ensemble mean. Therefore, a much stronger intensification is visible in the observation in contrast to the ensemble. The observation follows the intense cluster Intense (I) relatively closely up to a forecast time of 72 h. In contrast, over the same period, the weak cluster Weak (W) is not able to simulate the intensity adequately from the beginning, but especially after 36 h. This marks a critical point because the ensemble and the two clusters demonstrate an enhanced spread that indicates an increasing prediction uncertainty. Furthermore, the observation displays a stronger intensification between 36 and 48 h than both clusters. From 48 to 72 h, the intensification rate is again close to the mean value of cluster I. These two factors demonstrate the false prediction of a high number of members in the ensemble. After 72 h, neither cluster simulates the pressure change close to the observation, and also cluster I is unable to predict the increasing intensification rate. To identify the potential causes of the enhanced spread after 36 h, the causes are analyzed at this time through the study of several environmental conditions that are conducive to TCI. In contrast to the over-forecasting events, this storm indicates at a time where the cyclone's translation speed is reduced, and there, the system demonstrates an anticyclonic movement or a full loop, no increased forecast discrepancies (Fig.4.4). Champi demonstrates a straight track for all members at the critical point and a relatively high translation speed (Fig.6.2).

6.1.1 Dynamic parameters

The mean deep-layer environmental vertical wind shear at initial time for the cluster values is between 15.9 $\mathrm{m\,s^{-1}}$ and 16.8 $\mathrm{m\,s^{-1}}$. Cluster I shows a slight decrease after 6 h, while W remains constant. These high values are unfavorable for intensification of TCs (Goldenberg et al., 2001). However, cluster I intensifies after 36 h continuously, even though the values remain unfavorable for TCI. In contrast, W displays a drop after 36 h that continues up to 108 h forecast time. The spread in the different clusters rises after 36 h, especially for I; such divergence indicates insecurity in the forecast. This result is consistent with the findings of (Zhang and Tao, 2013). Furthermore,

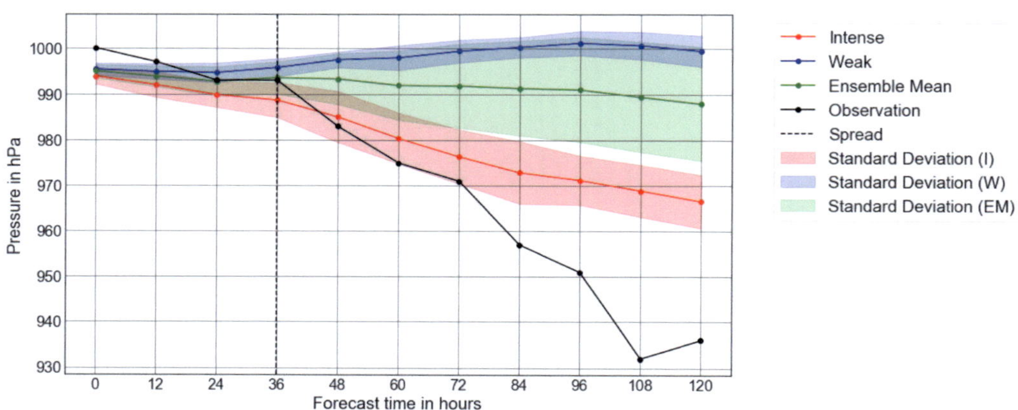

Figure 6.1: Time series of mean minimum pressure for Typhoon Champi for cluster I (red), cluster W (blue), observation (black), and ensemble mean (green). The corresponding standard deviations of clusters and the ensemble are shaded.

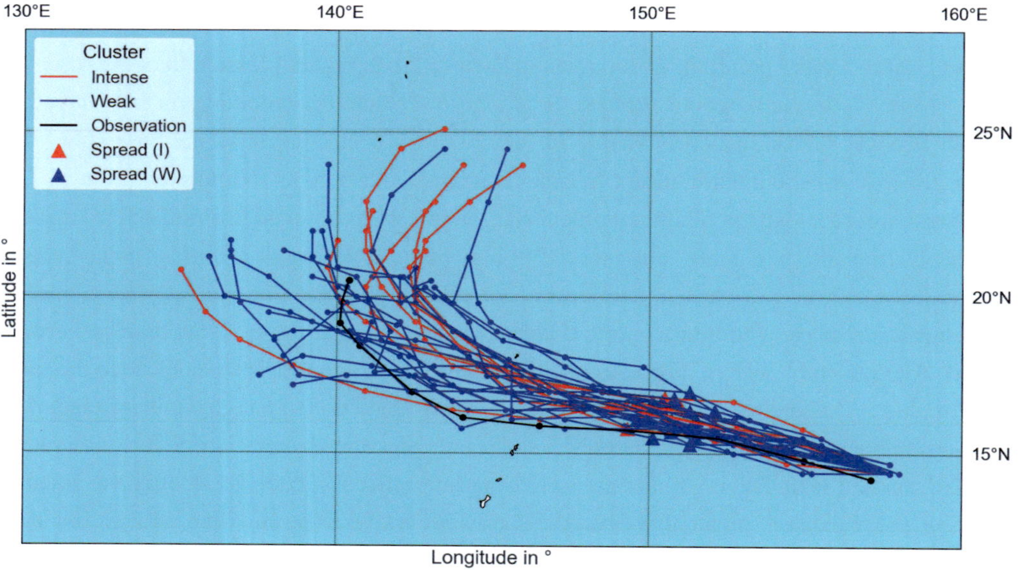

Figure 6.2: Tracks for I (red) and W (blue). Dots demonstrate (in 12 h time steps) the location of simulated cyclones and triangles the location at the critical point.

Emanuel and Nolan (2004) state that errors in the prediction of environmental vertical wind shear lead to large errors in the forecast of storm intensity. This result is confirmed in this case study: after 36 h, forecast time increases the ensemble spread of the vertical wind shear and pressure simultaneously. (Fig.6.3) illustrates that 31 % of the members hold values above 20 m s^{-1}. This strongly amplified wind field is related to strong westerlies in the upper troposphere, which are detrimental to TC development and intensification. The ensemble mean is close to the mean of W up to a forecast time of 36 h and lies between the two clusters after intensification. These results indicate that the environmental deep-layer wind shear is not contributory to the enhanced pressure fall after 36 h and especially not to the further intensification between 36 and 72 h forecast time. Therefore, it is critical to analyze the low- and mid-level moisture since earlier studies (Frank and

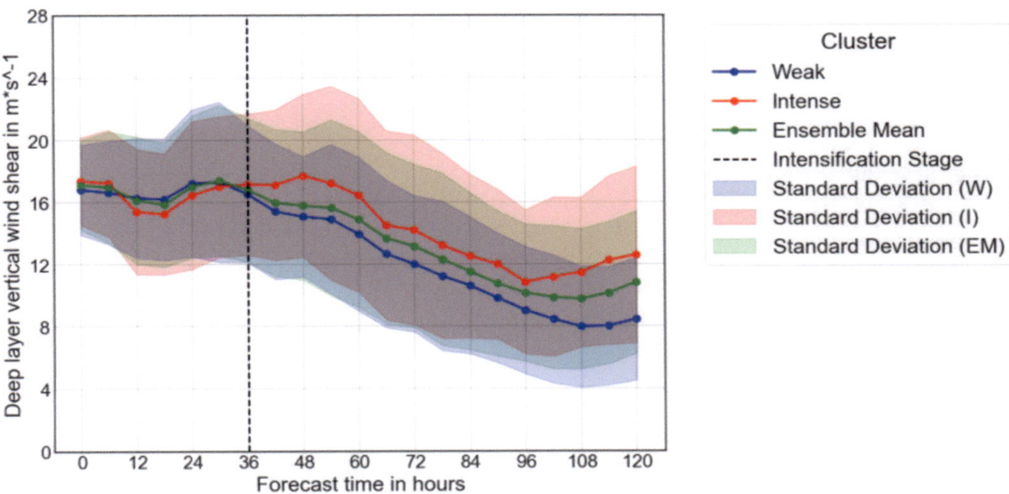

Figure 6.3: Time series of mean environmental vertical wind shear for Typhoon Champi for I (red), W (blue), and ensemble mean (green). The corresponding standard deviations are shaded. The environmental vertical wind shear is the mean value in the outer band of the storm in an annulus between 3° and 5° from the cyclone center.

Ritchie (1999); Komaromi and Majumdar (2014)) indicate that a TC structure is more persistent in environments with high moisture content.

6.1.2 Thermodynamic Parameters

Therefore, the relative humidity at 700 hPa is used as a proxy for the lower midlevel moisture. For both clusters, this moisture variable illustrates a layer with very high values in the western cyclone area (Fig.6.4). This layer is extended from the northern outside area over the western cyclone region to the southern part. At the storm system's northwest is a dry air layer with relative humidity under 40 %, which enhances the outward-directed moisture gradient and contributes to radiationally forced convergence (Gray, 1998). In contrast to W, cluster I indicates a higher humidity in the western part of the inner cyclone area, with values exceeding 80 %. At the same time, a relatively dry layer is visible to the west of the inner core. In comparison to W, this dry air is separated from a moist band around the cyclone area. The wind field at 700 hPa indicates that this high-humidity air is transported into the inner core of Champi. The normalized differences show that the eastern and northern outer bands are more humid for I than for W. The corresponding values are under -1. This is followed by a much higher relative humidity in the inner western core and the northern and western outer bands of the intense simulated cyclone. Due to the high vertical wind shear, the inner core at 700 hPa shifts to the west. The dry pattern to the west of the inner core is possible due to dry entrainments from the midlevel. According to Ge et al. (2013), the right quadrant of the down-shear side is a critical area for dry intrusions from the midlevel. Therefore, the western and northwestern areas of the simulated storms provide the inflow of low-entropy air into the storm center's boundary layer. However, in both cases, these areas are very moist, with a relative humidity above 80 % close to or in the western inner circle. This moist layer contributes to a higher resistance against the intruding low-entropy air, as confirmed by earlier studies. For

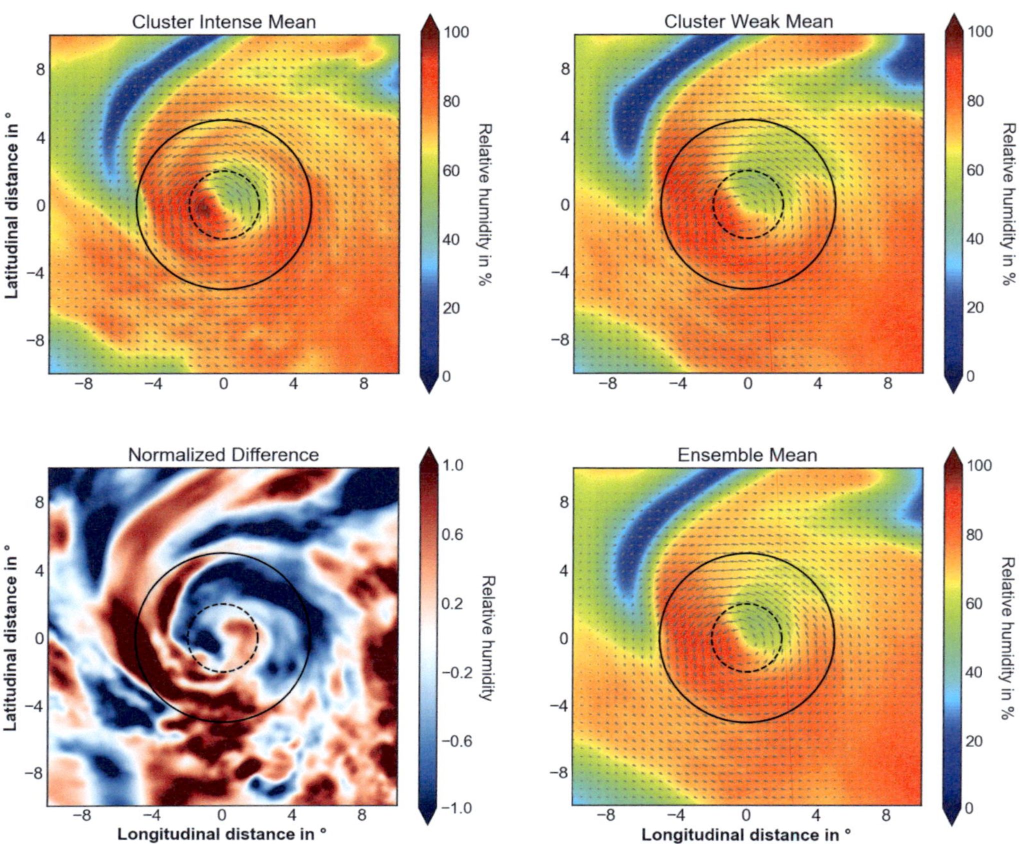

Figure 6.4: Storm-relative composites of mean relative humidity after 36 h forecast time at a pressure level of 700 hPa for I (top-left), W (top-right), and ensemble mean (bottom-right). A calculated normalized difference is shown at the bottom left. The wind field is indicated by black arrows at the corresponding pressure level. The black circle demonstrates the inner core of the cyclone.

example, Frank and Ritchie (1999) have shown through numerical studies that a vortex sustains its intensity much more effectively in an environment with high vertical wind shear if its surroundings are very moist.

This resistance is also illustrated by the boundary layer-specific humidity (Fig.6.5): both clusters and the ensemble mean demonstrate high specific humidity values above $18\,\mathrm{g\,kg^{-1}}$. Cluster I displays core values above $20\,\mathrm{g\,kg^{-1}}$, W about the core values above $20\,\mathrm{g\,kg^{-1}}$. Cluster W of about $18\,\mathrm{g\,kg^{-1}}$. Thus, the normalized difference is under -1. To the west and southwest, positive values close to 0.5 are recorded. Cluster I indicates a very moist restricted inner core and drier outer areas than cluster W. These are preferable conditions for a stronger convective activity that leads to a higher amount of latent heat release and hence to an increased pressure fall (McBride, 1995). The restricted area of convective heating results in higher efficiency in the pressure fall (Smith and Montgomery, 2016a). Additionally, Nolan et al. (2007) found that diabatic heating within

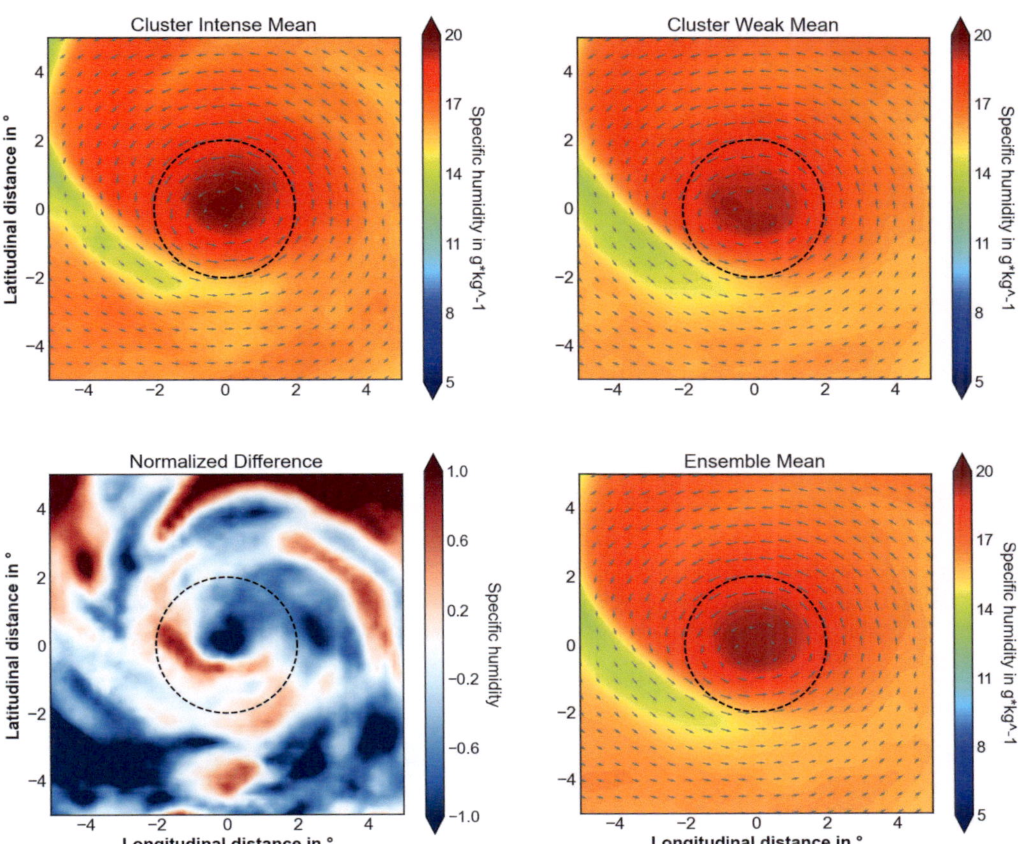

Figure 6.5: Storm-relative composites of mean relative humidity after 24 h forecast time at a pressure level of 925 hPa for I (top-left), W (top-right), and ensemble mean (bottom-right). A calculated normalized difference is shown at the bottom-left. The wind field is indicated by black arrows at the corresponding pressure level. The black circle demonstrates the inner core of the cyclone.

a circle with a radius of 100km from the center provides the most significant influence on TCI. Carrasco et al. (2014) investigated the influence of TC size on intensification rates, concluding that a more restricted convective activity at the beginning of intensification leads to an enhanced radiative difference in the higher troposphere, which forces the secondary circulation based on TC intensity (Wing et al., 2016).

6.1.3 Ventilation of low entropy air

(Sec.6.1.1) indicates an unfavorable high mean environmental vertical wind shear above $15\,\mathrm{m\,s^{-1}}$ for both clusters up to the intensification stage. This strong wind change with increasing height increased the probability of ventilation of dry air into the TC center (Tang and Emanuel, 2010, 2012). (Fig.6.6) demonstrates no clear differences in the entropy deficit for the clusters and the ensemble mean.

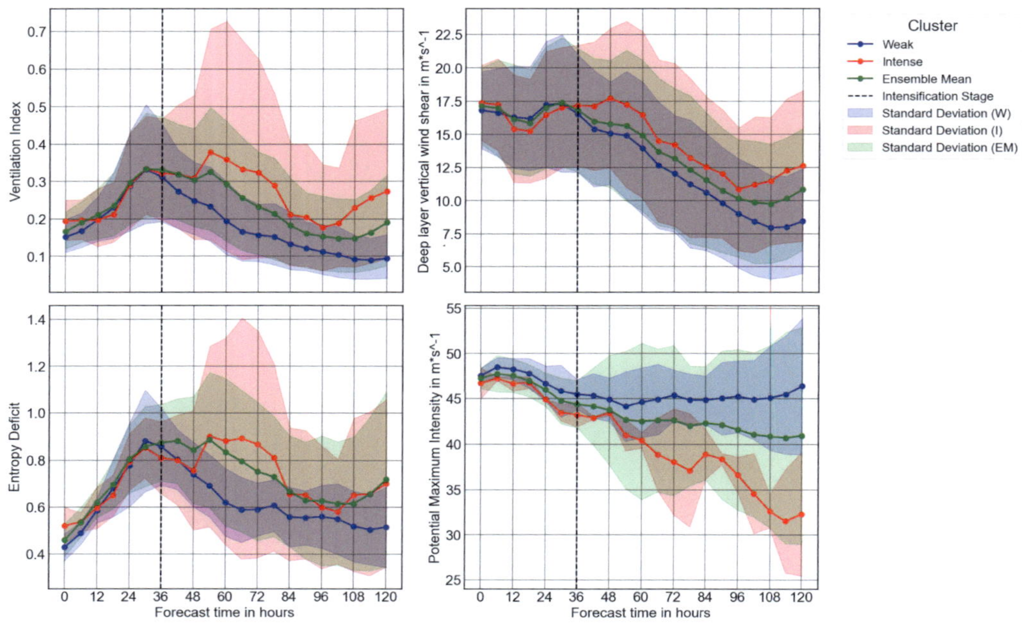

Figure 6.6: Mean values with forecast time of the ventilation index and its terms. The VI (top-left), deep layer vertical wind shear (top-right), entropy deficit (bottom-left), and potential maximum intensity (bottom-right). Dashed areas are the corresponding standard deviations.

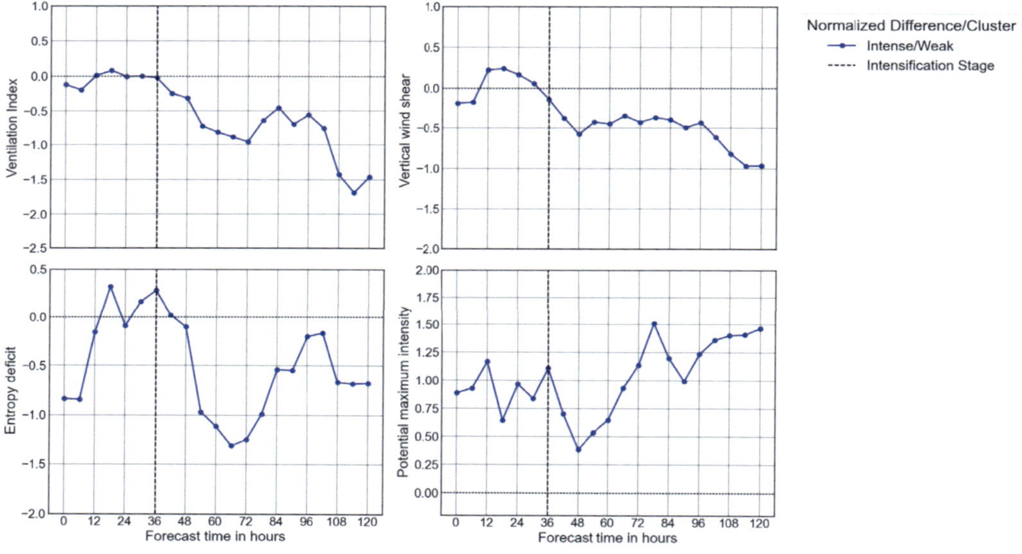

Figure 6.7: Time series of mean normalized differences over a circle with radius 2° from the storm center for the ventilation index (top-left), entropy deficit (bottom-left), potential maximum intensity (bottom-right), and vertical wind shear (top-right) in an outer circular area between 2° and 5° between W2 and I1. The dashed black line illustrates the first spread increase.

At 36h forecast time, a slightly higher entropy deficit is visible for W. This cluster possesses a maximum value of 0.88 at 30h, indicating unsuitable conditions for TCI. The values for both clusters decreased after intensification. However, at 48h, cluster I shows a sharp increase up to a mean value of 0.92. In addition, an enhanced spread is recognized, especially for I, which displays a high proportion (35.8 %) of values over 1. At the same time, the mean value of W is close to 0.6. These inconsistencies contribute to an enhanced vertical wind shear and a much higher ventilation index. Moreover, the potential maximum intensity is decreased after 48 h forecast time. This drop results from a lower sea surface temperature due to a more northward deflection of the storm tracks after 48 h for cluster I. The mean values for W for the whole period are between 44.2 $m\,s^{-1}$ and 48.7 $m\,s^{-1}$, while cluster I shows a decrease below 35 $m\,s^{-1}$.

Consequently, after 36 h forecast time, the ventilation index indicates an enhanced value for I in comparison to W. The largest discrepancies are recognizable between 54 and 72 h and at the end of the forecast period, where the normalized differences are under -2 (Fig.6.7). In addition, the mean values of both clusters up to 72 h forecast time are unsuited for an intensity increase. According to Tang and Emanuel (2012), a value of 0.1 is considered a threshold for intensification. Nevertheless, cluster intensity shows a further intensification during this period without any fluctuations, signaling that other processes are overcompensating for the detrimental effect of dry air intrusions. As shown in (Sec.6.1.2), the high amount of low-level moisture helps the TC to sustain the entrainment of low-entropy air from above. One reason for the low-level moisture is possibly the water vapor flux that provides the moisture via inward transport (Fritz and Wang, 2014). Thus, the water budget is analyzed in the following section to understand the source of this enhanced low-level moisture.

6.1.4 Water vapor fluxes

The VIWVF displays higher values for I over the whole cyclone area than for W, especially in the northern inner core area and outer bands, where the values are close to or less than -1 (Fig. 6.8). The highest values for all clusters and the ensemble mean are visible at the northwestern edge of the inner core. For I, the values are above 1,500 $kg\,m^{-2}$, while W only shows values close to 1,200 $kg\,m^{-2}$. In addition, cluster W indicates a much broader area with values close to zero. However, the negative outward-directed gradient is much more intense for I (Murthy and Boos, 2018). This negative gradient of surface enthalpy fluxes is necessary for TC development, as it provides greater convective instability near the circulation center. The intense simulated cyclone is characterized by a much stronger VIWVF, especially to the north in the inner core and to the north and east in the outer bands. However, the storm intensity between the clusters at 24 h forecast time is not exactly similar. Therefore, it is crucial to consider the outer area between 3 and 5°, which is less affected by the storm's wind field.

(Fig.6.9) illustrates that the mean normalized difference between W and I is -0.58 at 12 h forecast time. Up to 36 h before the intensification stage, this value decreases strongly. This drop continues up to 96 h, where the value exceeds -2. These negative values demonstrate the intense discrepancies between the clusters. The normalized difference of low-level divergence shows a neutral value at 12 h. However, it decreases up to 36 h forecast time to a value slightly above 0.1. The drop continues to 0.3 at 54 h. This minimum value indicates a slightly enhanced low-level convergence for I during the intensification stage compared to W. The upper-level divergence, meanwhile, is close to the enhanced spread and increase for cluster I. Therefore, the combination of much stronger VIWVF, slightly stronger low-level convergence, and upper-level divergence contributes to an intense

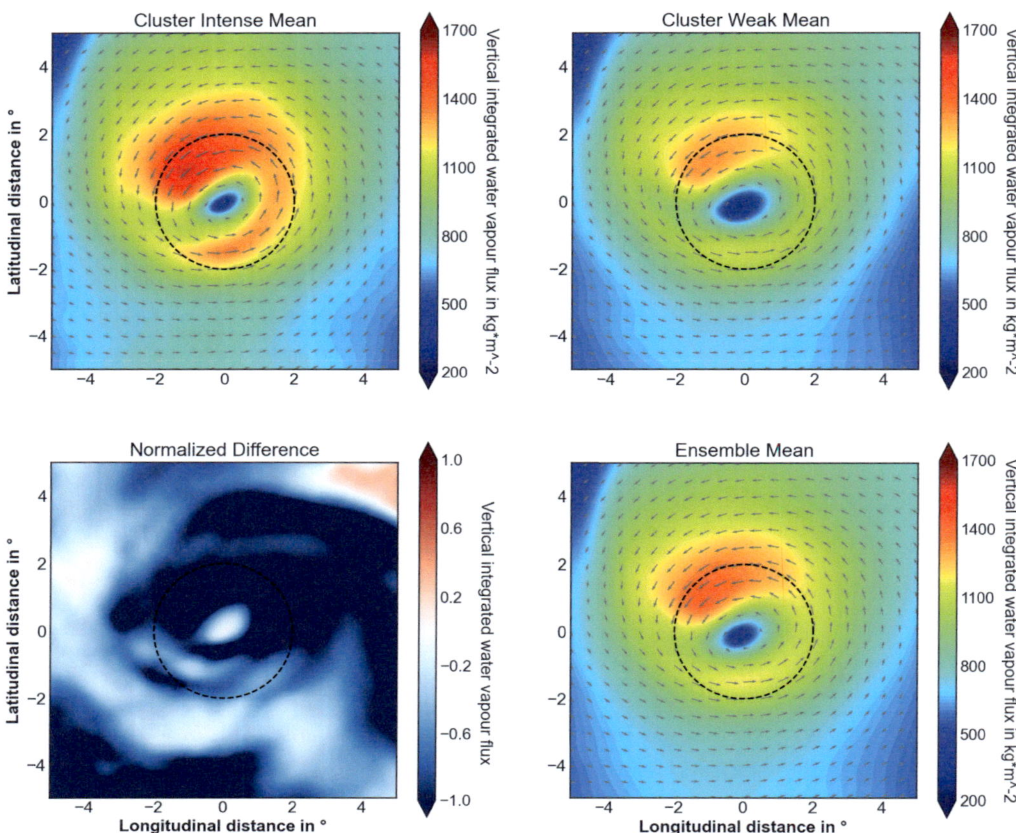

Figure 6.8: Storm-relative composites of mean VIWVF after 24 h forecast time for I (top-left), W (top-right), and ensemble mean (bottom-right). A calculated normalized difference is shown at the bottom left. The wind field is indicated by black arrows at a pressure level of 925 hPa. The black circle demonstrates the inner core of the cyclone.

drop in the normalized difference close to the beginning of the most powerful intensification rate. The powerful water vapor fluxes converge in the inner area of the cyclone and foster convective activity, resulting in a larger amount of latent heating and leading to increased precipitation rates and pressure fall in the storm center. This process forces the secondary circulation and hence contributes to positive feedback for further intensification.

Additionally, the moisture advection shows, especially in the southern inner core area, normalized differences under -1 (Fig.6.10). The wind field demonstrates negative normalized differences from the western area in the inner cyclone region. This band of low values indicates a much stronger moisture advection at 48 h forecast time for the simulated cyclone with an intensification than for the weak cyclone. This strong moisture advection in the upper branch of the boundary layer works against the dry entrainment from the midlevel (Sec.6.1.3) and contributes to the more substantial persistence of the intense cluster storms in contrast to the weak ones. A moist low layer is an

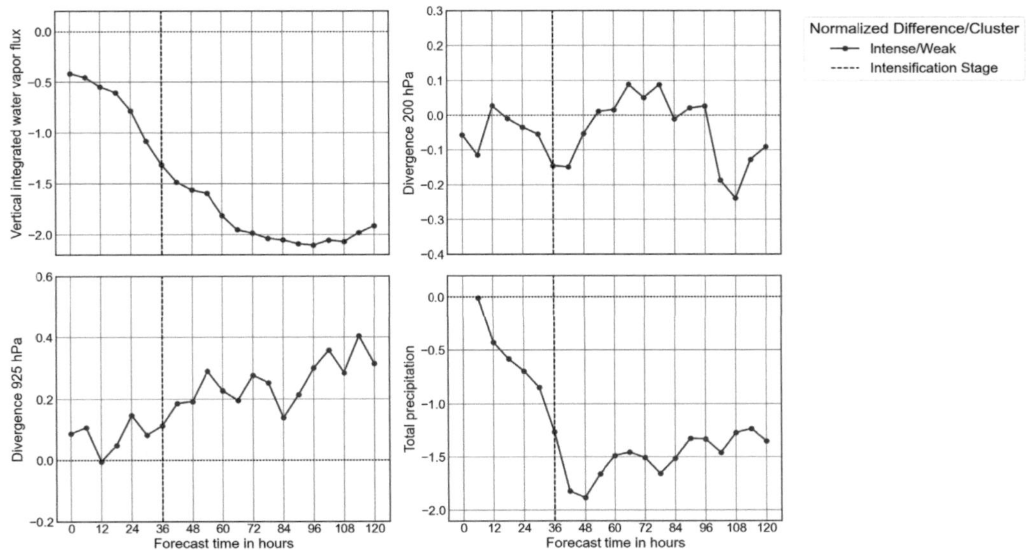

Figure 6.9: Time series of mean normalized differences over a circular area of 3° and 5° for the VIWVF (top-left), low-level divergence at 925 hPa (bottom-left), and upper-level divergence at 200 hPa (top-right). The mean normalized difference of total precipitation is averaged over the inner core. A circle with radius 2° from the center is defined as the inner core between W2 and I1. The dashed black line illustrates the first significant spread increase.

essential condition for TC development. (Wang and Hankes, 2016) found a nonlinear relationship between saturation fraction and precipitation rates. The normalized difference indicates after the critical point an intense rise of total precipitation for the intense cluster. This rise is associated with diabatic heating that fosters cyclone intensity. The following section compares this amount between the clusters and the different storms to quantify the contribution of diabatic heating to storm intensification.

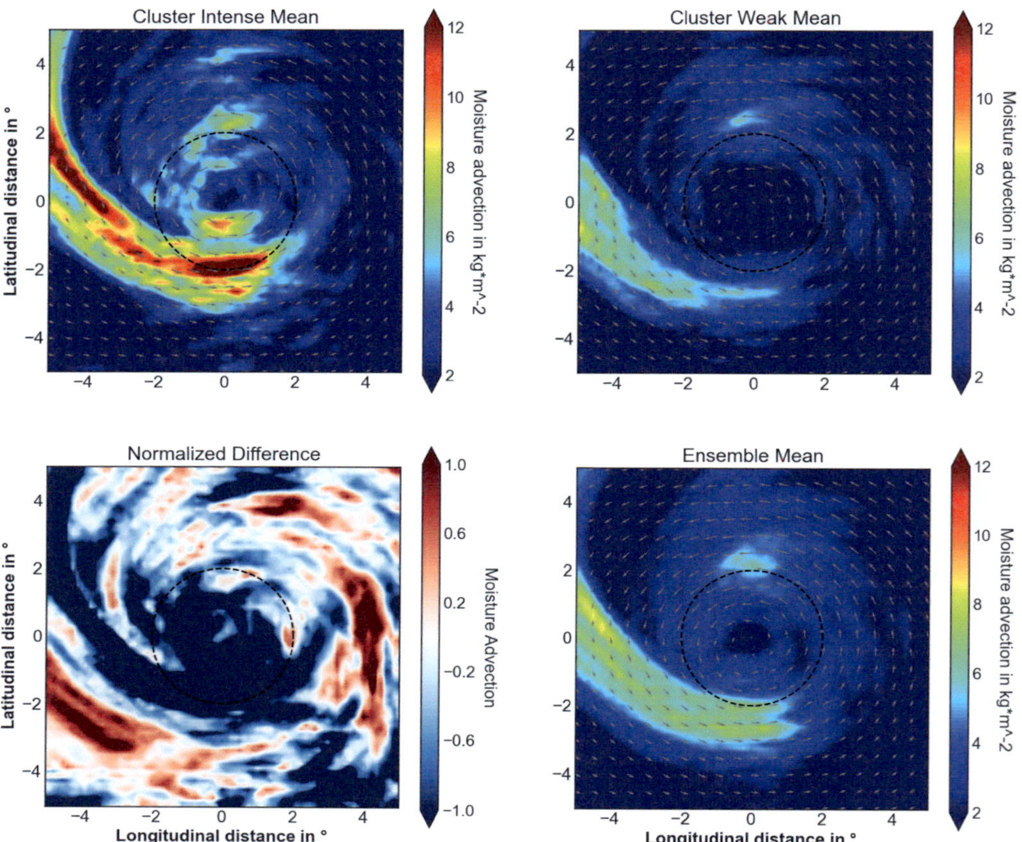

Figure 6.10: Storm-relative composites of mean moisture advection after 48 h forecast time for I (top-left), W (top-right), and all members (bottom-right). A calculated normalized difference is shown at the bottom-left. The wind field is indicated by black arrows at a pressure level of 925 hPa. Black circle demonstrates the inner core of the cyclone.

7 Diabatic Heating

An optimum choice of the upper boundary for the computation of the PTE terms is important for an accurate calculation (Sec.4.5). (Fig.7.1) illustrates that daily pressure fluctuations over 5 hPa are common in the lower and middle stratospheres. An insignificant layer of stratospheric dynamics is defined as a layer with 24 h pressure fluctuations under 2 hPa (Knippertz and Fink, 2008). Therefore, the layer at 100 hPa demonstrates significant dynamics over large parts of the WNP. In the middle stratosphere at 50 hPa, these fluctuations are higher than in the lower layer. Values locally exceed 10 hPa. This result is in contrast to (Knippertz and Fink, 2008), who show a decrease in the dynamics between the lower and middle stratosphere from maximum values of 15 hPa to 5 hPa. The maximum pressure fluctuations in the WNP during the case studies indicate at 100 hPa values of 10 hPa and at 50 hPa up to 25 hPa. These 24 h pressure changes decrease from the lower part to the higher part of the middle stratosphere (10 hPa) to maximum local values of 5 hPa. Furthermore, at 10 hPa, large parts of the WNP exhibit pressure fluctuations below 2 hPa. As a result, the calculation of the PTE terms is applied with an upper boundary of 10 hPa.

The PTE terms for Super Typhoon Champi demonstrate a substantial increase in vertical motion for cluster intensity after 6 h (Fig.7.2). Consequently, the diabatic heating term (computed from the residuum of the VMT and TADV terms) indicates a sharp rise in its contribution to pressure fall, with a value exceeding -40 hPa. The heating term shows the first minimum at 18 h, just after the intensification stage of Champi. In contrast, no strong increase was visible for the weak cluster. The diabatic term displays at the initial time a contribution of $-16,4$ hPa to pressure fall. This term remains nearly constant up to 12 h forecast time. Subsequently, the vertical motions of the intense cluster are much more powerful than the weak one. In comparison with findings from previous studies of the PTE for extratropical storms, the pressure fall values for the weak cluster with no intensification are close to intense extratropical winter storms. However, for the intense cluster, a three times higher heating amount to pressure fall is visible than Storm Klaus or Martin's maximum contribution, which showed the strongest contribution to pressure fall for the considered storms (Fink et al., 2012). These increased heating rates are a result of the much higher water vapor concentration in the tropics. This process is described by the Clausius Clapeyron equation (Koutsoyiannis, 2012). This equation describes how much water vapor a system provides for a defined temperature; for example, under standard pressure conditions, at a temperature of 25° C, the saturation vapor pressure is five times higher than at 0° C.

Diabatic heating is influenced by condensation and radiation processes. Due to the heavy precipitation in TCs, both clusters indicate a negative EP term. Consequently, this term demonstrates higher negative values with increasing heating rates. The TADV term for the intense cluster reports higher positive contributions, highlighting the increasingly cold air advection in the warm core of

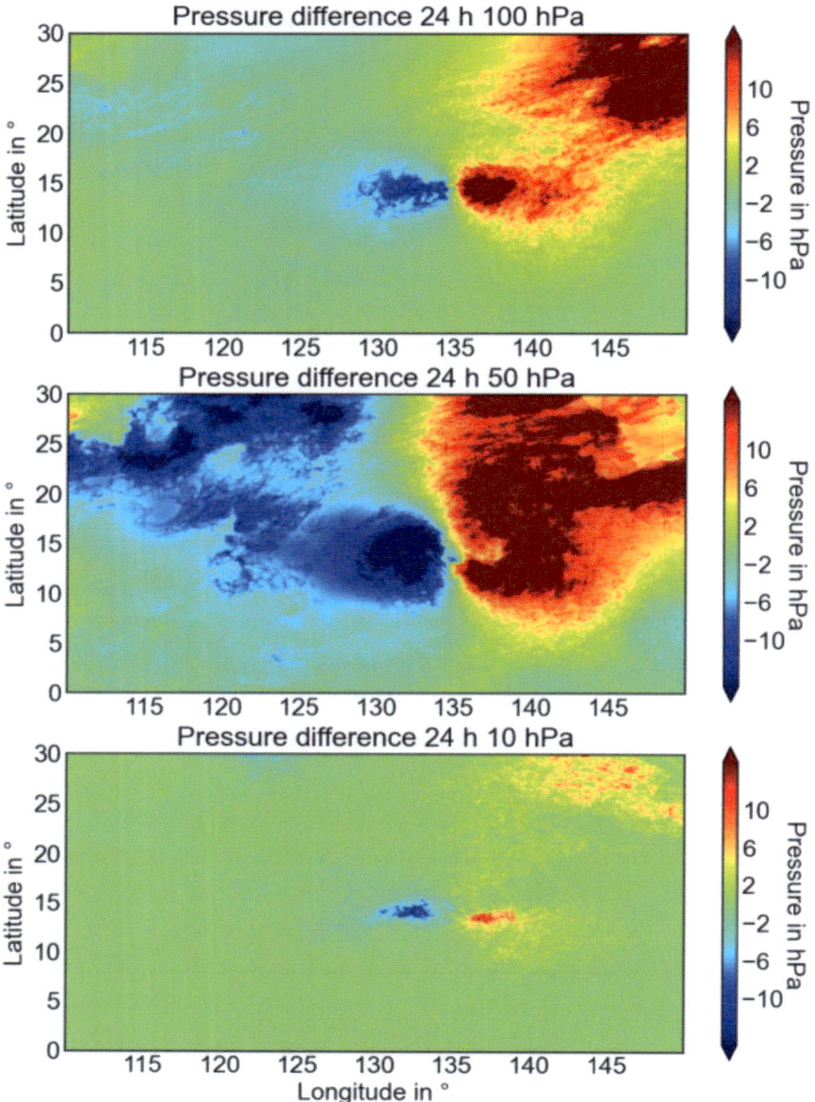

Figure 7.1: 24 h pressure change in different stratospheric layers over the WNP: at 100 hPa (top), 50 hPa (middle) and 10 hPa (bottom) on November 29, 2019, 12 UTC at 24 h forecast time for TC Kammuri.

the TC. As a result, the positive amount is more substantial for the intense cluster with a more intense cyclone that provides powerful subsidence in the inner storm center and, therefore, a warmer core. However, this amount exhibits no clear relationship with intensity. The most substantial contribution is visible at 12 h. At this time, the cyclone shows a much higher pressure than at 48 h forecast time. Therefore, the enhanced cold air advection is not only related to a warmer core. In comparison to Champi, TC Kammuri shows (especially at the beginning) a high fluctuating contribution. The intense cluster displays the first sustainable increase in diabatic heating after 36 h forecast time. At 36 h, the heating term demonstrates a maximum negative value of -61.4 hPa. The EP term rises from the beginning and displays a maximum at 48 h. Compared to the intense cluster,

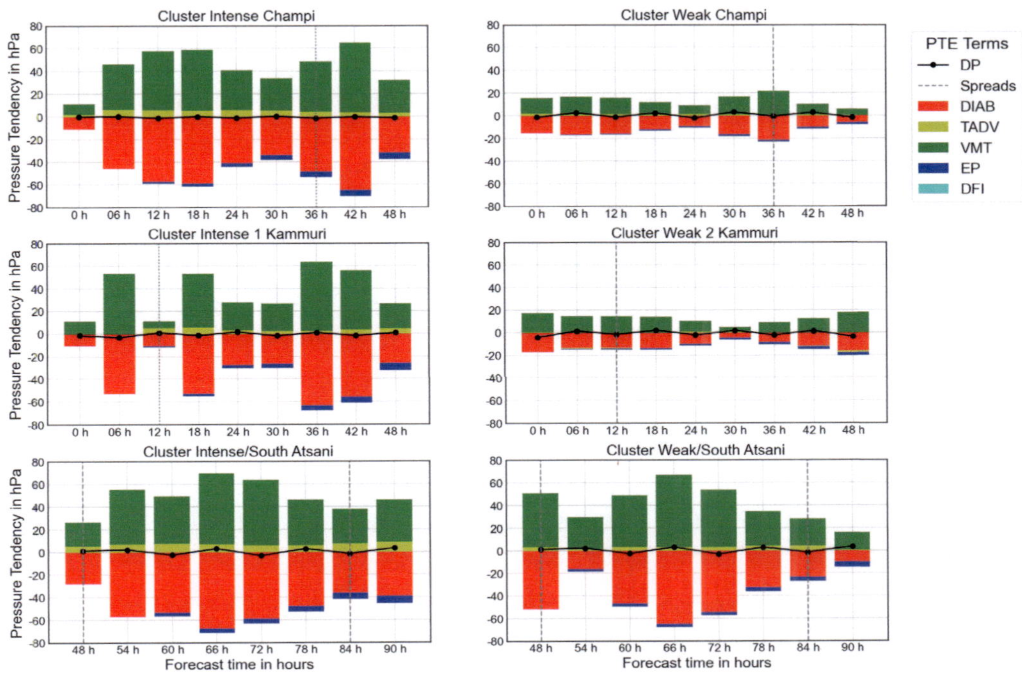

Figure 7.2: Contribution of PTE terms (DIAB (red), TADV (orange), VMT (green), EP (blue), DFI (cyan)) to pressure fall for TC Champi (top), Typhoon Kammuri (middle), and TC Atsani (bottom). The black line indicates the 6 h pressure tendency.

the weak cluster indicates a drop in diabatic heating up to 30 h forecast time. This decrease corresponds to the decreasing storm intensity during this period (Fig.5.19). A higher ventilation index leads to a drop in vertical motions and diabatic heating. At the same time, the fostered evaporation contributes to a minimal negative value of EP. After 30 h, the intensity remains constant, before the cluster indicates an intensification after 48 h. Consequently, the diabatic and vertical motion term contribution to pressure fall/rise increases in the period between 30 and 48 h. Additionally, the EP term shows an enhanced negative value that is associated with a higher precipitation efficiency. The TADV term is similar to Typhoon Champi. The intense clusters demonstrate a more intense cold advection into the center of the storm; this cold air transport contributes to higher amounts of TADV because a warmer core is one property of an intensified storm system (Stern and Nolan, 2012).

In contrast to the other storms, TC Atsani also displays high diabatic heating contributions (close to -60 hPa) in the weak cluster. These high values result from a stronger synoptic forcing due to the more northward location of Atsani. A trough to the northwest contributes to more powerful vertical motions close to the storm. Consequently, a higher negative value of the diabatic heating terms are illustrated. Additionally, a slight pressure decrease is indicated, especially in the period between 60 and 84 h forecast time. The strongest deviation between IS and WS is displayed just after the first intensification stage at 54 h and during the second intensification stage; such changes imply the important contribution of diabatic heating to pressure fall. As with the other storms, the TADV is higher for the intense cluster than for the weak cluster. Nonetheless, this positive

TADV value displays its maximum at 90 h, where the cyclone shows its most vigorous intensity. This distribution is in contrast to Typhoon Champi, where the TADV indicates no apparent relation to cyclone intensity. One possible reason for this is that the cold air advection into the center is compensated by warm, humid air advection from the outer bands into the inner region of the cyclone.

In sum, all the storms considered here demonstrate the valuable contribution of diabatic heating to TC intensification. TC Atsani, located more north and west than the other storms, indicates a much higher vertical motion for the weak cluster. These enhanced motions are a result of stronger synoptical forcing. A trough to the northwest not only influences the movement of the cyclone, but it also leads to more vigorous vertical motions. In addition, the storm's intensity displays a weak pressure drop over the considered period. The under-forecasting event provides no relation between storm intensity and the TADV term for the intense cluster, resulting in an enhanced cold air advection in the outer, inner area of Champi during the intensification stage. The total pressure change illustrates minor deviations (< 1.5 hPa) from the central pressure fall. This discrepancy results from the computation of mean values over an inner circle with radius $2°$ from the center. Additionally, the choice of the upper boundary is a possible error source because, at 10 hPa , local pressure fluctuations above 2 hPa are apparent over a period of 24 h.

8 Conclusion and Outlook

Tropical cyclones represent a severe threat to society and the economy. Erroneous predictions of TCs can contribute to failed evacuation measures and incorrect consulting of regional catastrophe managers, which are associated with high economic and socioeconomic costs (Bevere et al., 2020). Therefore, it is highly pertinent to improve the prediction skills of numerical weather prediction models. While storm track prediction capability has continuously increased in recent decades (Yamaguchi et al., 2017), no substantial improvement has been made in intensity forecasts (Rappaport et al., 2009; Cangialosi and Franklin, 2012; Bonavita et al., 2017). This is due to the fact that the complex interaction of dynamic and thermodynamic processes in tropical cyclones is still not fully understood. Previous investigations have indicated that these processes strongly influence predictive capability concerning tropical cyclone intensification (Komaromi and Majumdar, 2014; Zhang and Tao, 2013; Emanuel et al., 2004, TCI). Strongly amplified flow regimes with a high deep-layer vertical wind shear provide a high degree of sensitivity in ensemble forecasts. These regimes contribute to the uncertainty of vortex precession in numerical models that affect vortex spin up and the extension of dry intrusions from the midlevel (Zhang and Tao, 2013). A high deep-layer vertical wind shear is an inhibitor of intensification. This weakening process shows a higher probability in an environment with low moisture (Komaromi and Majumdar, 2015). Therefore, the water vapor budget equation defines the deep column moistening in the TC, which describes the moisture advection and the wind divergence.

This thesis investigates the potential causes of falsely predicted intensity forecasts of tropical cyclones (TCs) in the Western North Pacific using the ensemble prediction system of the European Center for Medium-Range Ensemble Forecasts (ECMWF). Three case studies are analyzed: one event with an under-predicted intensity forecast and two cases with over-forecasted intensity. A k-means clustering algorithm is applied to separate the ensemble members according to their intensity and location after a lead time with the strongest mean intensity error and a considerable spread in the ensemble. The focus here is on the study of sensitive acting environmental variables to intensity forecasts in the ensemble model. The ventilation index is computed to combine these variables and examine the terms with the most substantial impact on intensification. Additionally, a modified cold surge index is calculated to examine the impact of their strength on water vapor fluxes and pressure fall. The research questions defined in (Chap.3) can be answered as follows:

1. **How does the intensity of TCs evolve in the ensemble forecasts for increasing lead times?** All three storms show an increasing spread after a period of strong amplified stream fields with a high value of environmental deep-layer vertical wind shear. This result is consistent with previous studies (Zhang and Tao, 2013; Zhang and Sippel, 2009). Additionally, Munsell et al. (2013) states that miniscule and virtually unnoticeable discrepancies in the initial

conditions of a forecast dramatically determine the development of TCs. Furthermore, the rapidly increasing spread is associated with a lead time when the cyclone track demonstrates an anticyclonic looping or a turning and a speed decrease. This effect is visible for the two over-forecasting events, Atsani and Kammuri.

2. **What is the role of sensitive acting environmental variables on TCI?** The environmental deep-layer vertical wind shear only indicates a contribution to the increased pressure fall for the over-forecasting event Kammuri. For the under-forecasting event Champi and OFC TC Atsani, no apparent effect is recognizable. For Atsani and Champi, storm-relative composites for low-level core moisture exhibit a positive contribution to false TC intensity prediction. In contrast, the mid-level humidity at 700 hPa is only visible for Champi, a possible reason for the error afflicted intensity forecast. The other storms show no clear differences for mid-level humidity. Furthermore, Atsani and especially the UFC event Champi simulate a much stronger outward-directed moisture gradient for the intense cluster, which is most apparent in the lower midlevel at 700 hPa. Carrasco et al. (2014) have demonstrated that a TC with a higher intensification rate is defined over a smaller convective area close to the center with a drier outer area that is visible in the spatial moisture distribution in the lower and middle troposphere.

3. **Is there an influence from long-lasting northerly cold surges?** The case study results show that cold surge events lead to a higher intensification rate in the ensemble forecast of TC Atsani. This enhanced rate is only apparent for the strongest surges. For moderate and weak cold surges, no clear tendency is noticeable. The analyses of environmental variables indicate that the stronger cold surges lead to more intense vertically integrated water vapor fluxes (VIWVFs), which contribute to a positive feedback loop between moisture increase in the low-level core area and stronger secondary circulation that contributes to an intensity rise (Raymond and Kilroy, 2019).

4. **Does discrepancies in water vapor fluxes contribute to an overestimation or underestimation of TCI?** All three storms demonstrate a positive contribution of VIWVF to storm intensification. The most considerable effect is seen in the UFC event Champi. The water vapor fluxes lead to higher low-level moisture, which is an essential factor in intensification (Tao and Zhang, 2014). Champi and Kammuri both demonstrate (at the beginning, where only slight intensity differences are visible) a strong increase of horizontal and vertical water vapor fluxes in the inner cyclone core.

5. **Are there variances in the potential causes between individual case studies?** Under-forecasting event Champi illustrates, in comparison with the OFC events, higher discrepancies in the VIWVF between the intense and the weak cluster. Consequently, the low-level moisture differences are enhanced between the clusters. Kammuri indicates, in contrast to the more northward located OFC event Atsani and the UFC event Champi, that the environmental deep-layer vertical wind shear accounts for a high degree of the overprediction of TCI.

The answered research questions show that all over- and under-forecasting events demonstrate an influence of water vapor fluxes on discrepancies in intensification. Super Typhoon Champi shows the most intense difference between the intense and weak clusters. Before the simulated cyclones indicate a substantial discrepancy in intensification, the enhanced VIWWF associated with a more defined moisture advection into the TC inner core contributes to an effective low-level core moistening of the intense simulated storms. This deep column moistening is dependent on the water vapor budget equation, which includes a moisture advection term and a wind divergence term (Wu et al., 2013). The inward-directed, intense water vapor fluxes converge in the inner cyclone area; this convergence leads to vertical motions and air mass condensation. Champi demonstrates that the intense water vapor fluxes have their origin locally due to the enhanced air-sea enthalpy disequilibrium, which is consistent with the results from (Murthy and Boos, 2018). In contrast, Atsani exhibits the effect of an external cold surge on the strength of water vapor fluxes and the amount of total precipitation. This outcome is confirmed by (Fritz and Wang, 2014).

In turn, the pressure tendency equation indicates much higher contributions of diabatic heating to pressure fall in the intense cluster than in the weak cluster. However, this difference between the two clusters is much higher for Champi and Kammuri and much lower for TC Atsani. Higher humidity values diminish the detrimental effect of ventilation on the intense simulated storms in the boundary layer and the lower midlevel. Komaromi and Majumdar (2015) found that TCs also develop in the model by high deep-layer vertical wind shear of $15\,\mathrm{m\,s^{-1}}$ if the core of a developing storm system is very moist. This result is confirmed in this work: all three case studies indicate moderate or high environmental deep-layer vertical wind shear before the increasing forecast uncertainty. At the same time, the low-level moisture exhibits, especially for Champi and Atsani, humidity close to saturation. The highest moisture content is located on the down-shear side of the simulated storms. (Ge et al., 2013) identify the right quadrant of the vertical wind shear's down-shear direction as an area of the highest sensibility to dry air advection. This position is very moist, especially for typhoon Champi.

The ensemble predictions of Kammuri indicate a high spread of deep-layer vertical wind shear just before the intensification stage. This uncertainty is consistent with previous studies (Tao and Zhang (2014); Zhang and Tao (2013). However, this result is only valid for Kammuri. For TC Atsani and the under-forecasting event Champi, no apparent effects of the deep-layer environmental vertical wind shear on the error predicted intensity were noticeable. Although vertical wind shear is a dominant factor in the strength of ventilation, only Typhoon Kammuri indicates a positive contribution of a lower ventilation index to intensification; the other two storms demonstrate no such signal and, after intensification, the reverse. Atsani is the only storm that expresses favorable conditions for TC development for over 50 % of ensemble members in both clusters due to the ventilation index; in contrast, the other storms indicate unsuitable values over 0.1 mainly due to high values of environmental deep-layer vertical wind shear.

This thesis contributes to a better understanding of why the intensity ensemble prediction of tropical cyclones at the European Center for Medium-range Forecasts (ECMWF) is, in some cases, still highly erroneous. In particular, the k-means clustering and the analysis of sensitive acting environmental variables provide a deeper insight into the physical processes leading to intensity changes of cyclones. However, the complex physical interaction between environment and small-scale inner core dynamics is still not completely understood. Further research is thus required to gain a more specific comprehension of the small-scale dynamics of tropical storms.

In this work, the normalized difference for the VIWVF is calculated over a circular annulus between 3 and 5° from the TC center. This circular annulus is applied to consider the influence of the wind field on the strength of water vapor fluxes. However, the extension of this wind field varies between the simulated storms and with increasing forecast time. Therefore, it is crucial to contemplate the area between the individual simulated storms and forecast times separately. The radius of maximum winds could be used to identify extension changes in the wind field. Furthermore, the environmental relative humidity may be applied to determine the shape of the wind field (Hill and Lackmann, 2009). An extension of the case studies to other ocean basins, a higher sample size, and statistical significance testing could be of interest to examine the robustness of these results and compare findings between different ocean basins. One exception is the study of sub-seasonal cold surge influence on tropical storm intensity. This localized event should be studied over the SCS for a robust sample size of cyclones with a storm-relative consideration (Sec.4.4) and following the method according to (Yokoi et al., 2009). Likewise, it is necessary to extend the period for storm selection, although broadening the time range to years pre-2011 leads to poor comparability because of distinct model versions.

An extension to other ensemble forecasting model centers could provide a comparison between the individual ensemble forecast models of the centers. Such analysis would represent an important improvement in understanding the widespread misconceptions in ensemble forecasting models and would provide valuable knowledge to forecasters. Applied statistical post-processing calibration methods could correct the non-dispersive behavior in ensemble forecasts that lead to over-prediction and under-prediction of tropical cyclone intensity (Vannitsem et al., 2021). Maier-Gerber et al. (2021) used a statistical-dynamical approach to compare subregional North Atlantic TC occurrence forecasts over long periods up to week 5. This method could be extended to the WNP and other ocean basins. Additionally, it is important to consider the variation of the upper-level boundary when applying the PTE in different ocean basins. Thus, it is necessary to calculate pressure tendencies in different stratospheric layers to recognize the layer of insignificant dynamics (Knippertz and Fink, 2008). Furthermore, the use of diabatic tendencies limits the errors due to discretization for the residual computation of the diabatic heating term.

Bibliography

Bevere, L., M. Gloor, and A. Sobel, 2020: *Natural catastrophes in times of economic accumulation and climate change*. Swiss Re Institute.

Bister, M., and K. A. Emanuel, 2002: Low frequency variability of tropical cyclone potential intensity 1. interannual to interdecadal variability. *Journal of Geophysical Research: Atmospheres*, **107 (D24)**, ACL 26–1–ACL 26–15.

Bonavita, M., and Coauthors, 2017: On the initialization of tropical cyclones. *ECMWF Technical Memoranda*, **(810)**, URL https://www.ecmwf.int/node/17677.

Boyle, J. S., 1987: Synoptic aspects of the wintertime east asian monsoon. *Monsoon Meteorology*, 125–160, URL https://ci.nii.ac.jp/naid/10013453921/en/.

Braun, S. A., 2006: High-resolution simulation of hurricane bonnie (1998). part ii: Water budget. *Journal of the Atmospheric Sciences*, **63 (1)**, 43 – 64, URL https://journals.ametsoc.org/view/journals/atsc/63/1/jas3609.1.xml.

Bryan, G. H., 2008: On the computation of pseudoadiabatic entropy and equivalent potential temperature. *Monthly Weather Review*, **136 (12)**, 5239 – 5245.

Cangialosi, J. P., and J. L. Franklin, 2012: 2010 national hurricane center forecast verification report. *National Hurricane Center*.

Carrasco, C., C. Landsea, and Y.-L. Lin, 2014: The influence of tropical cyclone size on its intensification. *Weather and Forecasting*, **29**, 582–590.

Chan, J., 2005: The physics of tropical cyclone motion. *Annual Review of Fluid Mechanics*, **37**, 99–128, doi:10.1146/annurev.fluid.37.061903.175702.

Chan, J. C., and C. Li, 2004: The east asia winter monsoon. *East Asian Monsoon*, World Scientific, 54–106.

Chang, C.-P., P. A. Harr, and H.-J. Chen, 2005: Synoptic disturbances over the equatorial south china sea and western maritime continent during boreal winter. *Monthly Weather Review*, **133 (3)**, 489 – 503.

Chang, C.-P., C.-H. Liu, and H.-C. Kuo, 2003: Typhoon vamei: An equatorial tropical cyclone formation. *Geophysical Research Letters*, **30 (3)**.

Cinco, T. A., and Coauthors, 2016: Observed trends and impacts of tropical cyclones in the philippines. *International Journal of Climatology*, **36 (14)**, 4638–4650, URL https://rmets.onlinelibrary.wiley.com/doi/abs/10.1002/joc.4659.

Compo, G. P., G. N. Kiladis, and P. J. Webster, 1999: The horizontal and vertical structure of east asian winter monsoon pressure surges. *Quarterly Journal of the Royal Meteorological Society*, **125 (553)**, 29–54.

Dunkerton, T. J., M. T. Montgomery, and Z. Wang, 2009: Tropical cyclogenesis in a tropical wave critical layer: easterly waves. *Atmospheric Chemistry and Physics*, **9 (15)**, 5587–5646, URL https://acp.copernicus.org/articles/9/5587/2009/.

Emanuel, K., 1993: The physics of tropical cyclogenesis over the eastern pacific. *Tropical cyclone disasters*, **136**, 142.

Emanuel, K., 2003: Tropical cyclones. *Annual Review of Earth and Planetary Sciences*, **31 (1)**, 75–104.

Emanuel, K., C. DesAutels, C. Holloway, and R. Korty, 2004: Environmental control of tropical cyclone intensity. *Journal of the Atmospheric Sciences*, **61 (7)**, 843 – 858.

Emanuel, K., and D. S. Nolan, 2004: Tropical cyclone activity and the global climate system. *26th Conference on Hurricanes and Tropical Meteorolgy*.

Emanuel, K. A., 1986: An air-sea interaction theory for tropical cyclones. part i: Steady-state maintenance. *Journal of Atmospheric Sciences*, **43 (6)**, 585 – 605.

Fink, A. H., S. Pohle, J. G. Pinto, and P. Knippertz, 2012: Diagnosing the influence of diabatic processes on the explosive deepening of extratropical cyclones. *Geophysical Research Letters*, **39 (7)**.

Fink, A. H., and P. Speth, 1998: Tropical cyclones. *Naturwissenschaften*, URL https://doi.org/10.1007/s001140050536.

Frank, W. M., and E. A. Ritchie, 1999: Effects of environmental flow upon tropical cyclone structure. *Monthly weather review*, **127 (9)**, 2044–2061.

Frank, W. M., and E. A. Ritchie, 2001: Effects of vertical wind shear on the intensity and structure of numerically simulated hurricanes. *Monthly Weather Review*, **129 (9)**, 2249 – 2269.

Fritz, C., and Z. Wang, 2013: A numerical study of the impacts of dry air on tropical cyclone formation: A development case and a nondevelopment case. *Journal of the Atmospheric Sciences*, **70 (1)**, 91 – 111, URL https://journals.ametsoc.org/view/journals/atsc/70/1/jas-d-12-018.1.xml.

Fritz, C., and Z. Wang, 2014: Water vapor budget in a developing tropical cyclone and its implication for tropical cyclone formation. *Journal of the Atmospheric Sciences*, **71 (11)**, 4321 – 4332.

Fudeyasu, H., K. Ito, and Y. Miyamoto, 2018: Characteristics of tropical cyclone rapid intensification over the western north pacific. *Journal of Climate*, **31 (21)**, 8917 – 8930.

Fudeyasu, H., and R. Yoshida, 2018: Western North Pacific Tropical Cyclone Characteristics Stratified by Genesis Environment. *Monthly Weather Review*, **146 (2)**, 435–446.

Gao, S., S. Jia, Y. Wan, T. Li, S. Zhai, and X. Shen, 2019: The role of latent heat flux in tropical cyclogenesis over the western north pacific: Comparison of developing versus non-developing disturbances. *Journal of Marine Science and Engineering*, **7 (2)**, 28.

Ge, X., T. Li, and M. Peng, 2013: Effects of vertical shears and midlevel dry air on tropical cyclone developments. *Journal of the Atmospheric Sciences*, **70 (12)**, 3859 – 3875.

Ge, X., T. Li, Y. Wang, and M. Peng, 2008: Tropical cyclone energy dispersion in a three-dimensional primitive equation model: Upper tropospheric influence. *Journal of the Atmospheric Sciences*, **65**, 2272–2289.

Goldenberg, S. B., C. W. Landsea, A. M. Mestas-Nuñez, and W. M. Gray, 2001: The recent increase in atlantic hurricane activity: Causes and implications. *Science*, **293 (5529)**, 474–479.

Gray, W. M., 1968: Global view of the origin of tropical disturbances and storms. *Monthly Weather Review*, **96 (10)**, 669–700.

Gray, W. M., 1975: Tropical cyclone genesis. Ph.D. thesis, Colorado State University. Libraries.

Gray, W. M., 1977: Tropical cyclone genesis in the western north pacific. *Journal of the Meteorological Society of Japan. Ser. II*, **55 (5)**, 465–482.

Gray, W. M., 1998: The formation of tropical cyclones. *Meteorology and Atmospheric Physics*, **67**, 37–69.

Hamill, T. M., J. S. Whitaker, M. Fiorino, and S. G. Benjamin, 2011: Global ensemble predictions of 2009's tropical cyclones initialized with an ensemble kalman filter. *Monthly Weather Review*, **139 (2)**, 668 – 688.

Hartigan, J. A., and M. A. Wong, 1979: Algorithm as 136: A k-means clustering algorithm. *Journal of the Royal Statistical Society. Series C (Applied Statistics)*, **28 (1)**, 100–108, URL http://www.jstor.org/stable/2346830.

Helms, C. N., and R. E. Hart, 2015: The evolution of dropsonde-derived kinematic and thermodynamic structures in developing and nondeveloping atlantic tropical convective systems. *Monthly Weather Review*, **143 (8)**, 3109 – 3135, URL https://journals.ametsoc.org/view/journals/mwre/143/8/mwr-d-14-00242.1.xml.

Hill, K. A., and G. M. Lackmann, 2009: Influence of environmental humidity on tropical cyclone size. *Monthly Weather Review*, **137 (10)**, 3294 – 3315, URL https://journals.ametsoc.org/view/journals/mwre/137/10/2009mwr2679.1.xml.

Houze, R. A., S. G. Geotis, F. D. Marks, and A. K. West, 1981: Winter monsoon convection in the vicinity of north borneo. part i: Structure and time variation of the clouds and precipitation. *Monthly Weather Review*, **109 (8)**, 1595 – 1614, URL https://journals.ametsoc.org/view/journals/mwre/109/8/1520-0493_1981_109_1595_wmcitv_2_0_co_2.xml.

IBTRACS, 2015: 2015 super typhoon champi (2015286n13161). URL http://ibtracs.unca.edu/index.php?name=v04r00-2015286N13161, acessed: 2021-10-27.

IBTRACS, 2019: 2019 super typhoon kammuri (2019329n09160). URL http://www.atms.unca.edu/ibtracs/ibtracs_v04r00/index.php?name=v04r00-2020304N08148, acessed: 2021-11-28.

IBTRACS, 2020: 2020 severe tropical storm atsani (2020304n08148). URL http://ibtracs.unca.edu/index.php?name=v04r00-2020304N08149, acessed: 2021-10-27.

Knippertz, P., and A. H. Fink, 2008: Dry-season precipitation in tropical west africa and its relation to forcing from the extratropics. *Monthly Weather Review*, **136 (9)**, 3579 – 3596, URL https://journals.ametsoc.org/view/journals/mwre/136/9/2008mwr2295.1.xml.

Komaromi, W. A., and S. J. Majumdar, 2014: Ensemble-Based Error and Predictability Metrics Associated with Tropical Cyclogenesis. Part I: Basinwide Perspective. *Monthly Weather Review*, **142 (8)**, 2879–2898.

Komaromi, W. A., and S. J. Majumdar, 2015: Ensemble-Based Error and Predictability Metrics Associated with Tropical Cyclogenesis. Part II: Wave-Relative Framework. *Monthly Weather Review*, **143 (5)**, 1665–1686.

Koutsoyiannis, D., 2012: Clausius–clapeyron equation and saturation vapour pressure: simple theory reconciled with practice. *European Journal of physics*, **33 (2)**, 295.

Kumpf, A., B. Tost, M. Baumgart, M. Riemer, R. Westermann, and M. Rautenhaus, 2018: Visualizing confidence in cluster-based ensemble weather forecast analyses. *IEEE Transactions on Visualization and Computer Graphics*, **24 (1)**, 109–119.

Lagmay, A. M. F., and Coauthors, 2015: Devastating storm surges of typhoon haiyan. *International Journal of Disaster Risk Reduction*, **11**, 1–12, URL https://www.sciencedirect.com/science/article/pii/S2212420914000922.

Leutbecher, M., and T. N. Palmer, 2008: Ensemble forecasting. *Journal of computational physics*, **227 (7)**, 3515–3539.

Li, T., and P.-c. Hsu, 2018: Tropical cyclone formation. *Fundamentals of Tropical Climate Dynamics*, Springer, 107–147.

Liang, B., 1991: Tropical atmospheric circulation system over the south china sea. *China Meteorol. Press, Beijing*.

Longshore, D., 2008: *Encyclopedia of hurricanes, typhoons, and cyclones*. Infobase Publishing.

Lussier III, L. L., M. T. Montgomery, and M. M. Bell, 2014: The genesis of typhoon nuri as observed during the tropical cyclone structure 2008 (tcs-08) field experiment – part 3: Dynamics of low-level spin-up during the genesis. *Atmospheric Chemistry and Physics*, **14 (16)**, 8795–8812, URL https://acp.copernicus.org/articles/14/8795/2014/.

Maier-Gerber, M., A. H. Fink, M. Riemer, E. Schoemer, C. Fischer, and B. Schulz, 2021: Statistical–dynamical forecasting of subseasonal north atlantic tropical cyclone occurrence. *Weather and Forecasting*, **36 (6)**, 2127 – 2142, URL https://journals.ametsoc.org/view/journals/wefo/36/6/WAF-D-21-0020.1.xml.

McBride, J., 1995: Global perspectives on tropical cyclones. *World Meteorological Organization Report No. TCP-38*, **289**.

Munsell, E. B., F. Zhang, and D. P. Stern, 2013: Predictability and dynamics of a nonintensifying tropical storm: Erika (2009). *Journal of the Atmospheric Sciences*, **70 (8)**, 2505 – 2524, URL https://journals.ametsoc.org/view/journals/atsc/70/8/jas-d-12-0243.1.xml.

Murthy, V. S., and W. R. Boos, 2018: Role of surface enthalpy fluxes in idealized simulations of tropical depression spinup. *Journal of the Atmospheric Sciences*, **75 (6)**, 1811 – 1831.

Nguyen, K.-A., Y.-A. Liou, and J. P. Terry, 2019: Vulnerability of vietnam to typhoons: A spatial assessment based on hazards, exposure and adaptive capacity. *Science of the Total Environment*, **682**, 31–46.

Nolan, D. S., E. D. Rappin, and K. A. Emanuel, 2007: Tropical cyclogenesis sensitivity to environmental parameters in radiative–convective equilibrium. *Quarterly Journal of the Royal Meteorological Society*, **133 (629)**, 2085–2107, URL https://rmets.onlinelibrary.wiley.com/doi/abs/10.1002/qj.170.

Palmer, T., and Coauthors, 2007: The ensemble prediction system–recent and ongoing developments. *ECMWF Technical Memoranda*, **(430)**, 53.

Papavasileiou, G., A. Voigt, and P. Knippertz, 2020: The role of observed cloud-radiative anomalies for the dynamics of the north atlantic oscillation on synoptic time-scales. *Quarterly Journal of the Royal Meteorological Society*, **146 (729)**, 1822–1841.

Pinto, J. G., T. Spangehl, U. Ulbrich, and P. Speth, 2005: Sensitivities of a cyclone detection and tracking algorithm: individual tracks and climatology. *Meteorologische Zeitschrift*, **14 (6)**, 823–838.

Rappaport, E. N., and Coauthors, 2009: Advances and challenges at the national hurricane center. *Weather and Forecasting*, **24 (2)**, 395 – 419.

Raymond, D. J., and G. Kilroy, 2019: Control of convection in high-resolution simulations of tropical cyclogenesis. *Journal of Advances in Modeling Earth Systems*, **11 (6)**, 1582–1599.

Raymond, D. J., and C. López Carrillo, 2011: The vorticity budget of developing typhoon nuri (2008). *Atmospheric Chemistry and Physics*, **11 (1)**, 147–163, URL https://acp.copernicus.org/articles/11/147/2011/.

Riehl, H., 1948: On the formation of typhoons. *Journal of Atmospheric Sciences*, **5 (6)**, 247 – 265.

Riemer, M., M. T. Montgomery, and M. E. Nicholls, 2010: A new paradigm for intensity modification of tropical cyclones: thermodynamic impact of vertical wind shear on the inflow layer. *Atmospheric Chemistry and Physics*, **10 (7)**, 3163–3188.

Ritchie, E. A., and G. J. Holland, 1999: Large-scale patterns associated with tropical cyclogenesis in the western pacific. *Monthly Weather Review*, **127 (9)**, 2027 – 2043, URL https://journals.ametsoc.org/view/journals/mwre/127/9/1520-0493_1999_127_2027_lspawt_2.0.co_2.xml.

Russell, J. O., A. Aiyyer, J. D. White, and W. Hannah, 2017: Revisiting the connection between african easterly waves and atlantic tropical cyclogenesis. *Geophysical Research Letters*, **44 (1)**, 587–595.

Schubert, W. H., and J. J. Hack, 1982: Inertial stability and tropical cyclone development. *Journal of Atmospheric Sciences*, **39 (8)**, 1687 – 1697, URL https://journals.ametsoc.org/view/journals/atsc/39/8/1520-0469_1982_039_1687_isatcd_2_0_co_2.xml.

Shu, S., Y. Wang, and L. Bai, 2013: Insight into the role of lower-layer vertical wind shear in tropical cyclone intensification over the western north pacific. *Acta meteorologica sinica*, **27 (3)**, 356–363.

Smith, R. K., and M. T. Montgomery, 2016a: The efficiency of diabatic heating and tropical cyclone intensification. *Quarterly Journal of the Royal Meteorological Society*, **142 (698)**, 2081–2086.

Smith, R. K., and M. T. Montgomery, 2016b: Understanding hurricanes. *Weather*, **71 (9)**, 219–223, URL https://rmets.onlinelibrary.wiley.com/doi/abs/10.1002/wea.2776.

Stern, D. P., and D. S. Nolan, 2012: On the height of the warm core in tropical cyclones. *Journal of the Atmospheric Sciences*, **69 (5)**, 1657–1680.

Swinbank, R., and Coauthors, 2016: The TIGGE Project and Its Achievements. *Bulletin of the American Meteorological Society*, **97 (1)**, 49–67.

Takahashi, H. G., Y. Fukutomi, and J. Matsumoto, 2011: The impact of long-lasting northerly surges of the east asian winter monsoon on tropical cyclogenesi. *Journal of the Meteorological Society of Japan*.

Takakura, T., R. Kawamura, T. Kawano, K. Ichiyanagi, M. Tanoue, and K. Yoshimura, 2018: An estimation of water origins in the vicinity of a tropical cyclone's center and associated dynamic processes. *Climate Dynamics*, **50**, 1–15.

Tang, B., and K. Emanuel, 2010: Midlevel ventilations constraint on tropical cyclone intensity. *Journal of the Atmospheric Sciences*, **67 (6)**, 1817 – 1830, URL https://journals.ametsoc.org/view/journals/atsc/67/6/2010jas3318.1.xml.

Tang, B., and K. Emanuel, 2012: A ventilation index for tropical cyclones. *Bulletin of the American Meteorological Society*, **93 (12)**, 1901 – 1912.

Tao, D., and F. Zhang, 2014: Effect of environmental shear, sea-surface temperature, and ambient moisture on the formation and predictability of tropical cyclones: An ensemble-mean perspective. *Journal of Advances in Modeling Earth Systems*, **6**.

Torn, R. D., J. S. Whitaker, P. Pegion, T. M. Hamill, and G. J. Hakim, 2015: Diagnosis of the source of gfs medium-range track errors in hurricane sandy (2012). *Monthly Weather Review*, **143 (1)**, 132 – 152.

Van der Grijn, G., J. Paulsen, F. Lalaurette, and M. Leutbecher, 2005: Early medium-range forecasts of tropical cyclones. *ECMWF newsletter*, **102**, 7–14.

Van Sang, N., R. K. Smith, and M. T. Montgomery, 2008: Tropical-cyclone intensification and predictability in three dimensions. *Quarterly Journal of the Royal Meteorological Society*, **134 (632)**, 563–582, URL https://rmets.onlinelibrary.wiley.com/doi/abs/10.1002/qj.235.

Vannitsem, S., and Coauthors, 2021: Statistical postprocessing for weather forecasts: Review, challenges, and avenues in a big data world. *Bulletin of the American Meteorological Society*, **102 (3)**, E681 – E699, URL https://journals.ametsoc.org/view/journals/bams/102/3/BAMS-D-19-0308.1.xml.

Vigh, J. L., and W. H. Schubert, 2009: Rapid development of the tropical cyclone warm core. *Journal of the Atmospheric Sciences*, **66 (11)**, 3335 – 3350, URL https://journals.ametsoc.org/view/journals/atsc/66/11/2009jas3092.1.xml.

Vitart, F., F. Prates, A. Bonet, and C. Sahin, 2012: New tropical cyclone products on the web. *Ecmwf newsletter*, **130**, 17–23.

Wang, G., J. Su, Y. Ding, and D. Chen, 2007: Tropical cyclone genesis over the south china sea. *Journal of Marine Systems*, **68 (3)**, 318 – 326.

Wang, S., and R. Toumi, 2019: Impact of dry midlevel air on the tropical cyclone outer circulation. *Journal of the Atmospheric Sciences*, **76**.

Wang, Z., 2012: Thermodynamic aspects of tropical cyclone formation. *Journal of the Atmospheric Sciences*, **69 (8)**, 2433 – 2451, URL https://journals.ametsoc.org/view/journals/atsc/69/8/jas-d-11-0298.1.xml.

Wang, Z., and I. Hankes, 2016: Moisture and precipitation evolution during tropical cyclone formation as revealed by the ssm/i–ssmis retrievals. *Journal of the Atmospheric Sciences*, **73 (7)**, 2773 – 2781, URL https://journals.ametsoc.org/view/journals/atsc/73/7/jas-d-15-0306.1.xml.

Wing, A. A., S. J. Camargo, and A. H. Sobel, 2016: Role of radiative–convective feedbacks in spontaneous tropical cyclogenesis in idealized numerical simulations. *Journal of the Atmospheric Sciences*, **73 (7)**, 2633 – 2642, URL https://journals.ametsoc.org/view/journals/atsc/73/7/jas-d-15-0380.1.xml.

Wong, M. L. M., and J. C. L. Chan, 2004: Tropical cyclone intensity in vertical wind shear. *Journal of the Atmospheric Sciences*, **61 (15)**, 1859 – 1876.

Wu, W., C. Jilong, and R. Huang, 2013: Water budgets of tropical cyclones: Three case studies. *Advances in Atmospheric Sciences*, **30**.

Xu, D., C. Qiu, and Y. Yan, 2019: Role of the south china sea summer upwelling in tropical cyclone intensity. *American Journal of Climate Change*, **08**, 1–13.

Yamaguchi, M., J. Ishida, H. Sato, and M. Nakagawa, 2017: Wgne intercomparison of tropical cyclone forecasts by operational nwp models: A quarter century and beyond. *Bulletin of the American Meteorological Society*, **98 (11)**, 2337 – 2349.

Yang, M.-J., S. A. Braun, and D.-S. Chen, 2011: Water budget of typhoon nari (2001). *Monthly Weather Review*, **139 (12)**, 3809–3828.

Yokoi, S., Y. Takayabu, and J. Chan, 2009: Tropical cyclone genesis frequency over the western north pacific simulated in medium-resolution coupled general circulation models. *Climate Dynamics*, **33**, 665–683.

Yoshida, R., and H. Ishikawa, 2013: Environmental factors contributing to tropical cyclone genesis over the western north pacific. *Monthly Weather Review*, **141 (2)**, 451–467.

Yuan, J., T. Li, and D. Wang, 2015: Precursor synoptic-scale disturbances associated with tropical cyclogenesis in the south china sea during 2000–2011. *International Journal of Climatology*, **35 (12)**, 3454–3470, URL https://rmets.onlinelibrary.wiley.com/doi/abs/10.1002/joc.4219.

Zehr, R. M., 1992: Tropical cyclogenesis in the western north pacific.

Zhang, F., and J. A. Sippel, 2009: Effects of moist convection on hurricane predictability. *Journal of the Atmospheric Sciences*, **66 (7)**, 1944 – 1961.

Zhang, F., and D. Tao, 2013: Effects of vertical wind shear on the predictability of tropical cyclones. *Journal of the Atmospheric Sciences*, **70 (3)**, 975 – 983.

8.1 List of Abbreviations

SCS South China Sea 1

WNP Western North Pacific 1

TC Tropical cyclone 1

EPS Ensemble Prediction System 2

OFC over-forecasting 2

UFC under-forecasting 2

THORPEX The Observing system Research and Predictability Experiment 2

TIGGE THORPEX Interactive Grand Global Ensemble 2

PTE pressure tendency equation 3

TCI tropical cyclone intensification 1

TCG tropical cyclone genesis 6

IS Intense South 39

WS Weak South 39

CS Cold surge 47

NOCS No cold surge 47

VIWVF Vertically Integrated Water Vapor Flux 53